Industrial Automation: Systems and Engineering

Industrial Automation: Systems and Engineering

Geoffrey Williamson

STATES
ACADEMIC PRESS
www.statesacademicpress.com

Industrial Automation: Systems and Engineering
Geoffrey Williamson
ISBN: 978-1-63989-285-3 (Hardback)

STATES
ACADEMIC PRESS

Published by States Academic Press,
109 South 5th Street,
Brooklyn, NY 11249, USA

Cataloging-in-Publication Data

Industrial automation : systems and engineering / Geoffrey Williamson.
p. cm.
Includes bibliographical references and index.
ISBN 978-1-63989-285-3
1. Automation. 2. Industrial engineering. 3. Automatic control.
4. Automatic machinery. I. Williamson, Geoffrey.
T59.5 .I53 2022
670.427--dc23

For more information regarding States Academic Press and its products, please visit the publisher's website www.statesacademicpress.com

Table of Contents

Preface

This book has been written, keeping in view that students want more practical information. Thus, my aim has been to make it as comprehensive as possible for the readers. I would like to extend my thanks to my family and co-workers for their knowledge, support and encouragement all along.

Industrial automation is the technology which uses diverse control systems for handling different industrial processes and machineries with minimal human assistance. It facilitates production by increasing product quality, reliability, production rate and decreasing human error. It provides optimum cost of operation as the need of labor gets reduced. This field aims at replacing human decision-making and manual command-response activities with logical programming commands and mechanized equipment. Industrial robotics is a sub-branch of industrial automation. Industrial robots are the automated robot systems used in manufacturing processes. The use of these robots increases the safety level as it replaces personnel with machines in hazardous working conditions. Emerging technologies include automated mining, logistics automation and programmable logic controllers. The topics covered in this extensive book deal with the core aspects of industrial automation. The various sub-fields along with technological progress that have future implications are glanced at in it. This book will provide comprehensive knowledge to the readers.

A brief description of the chapters is provided below for further understanding:

Chapter - What is Automation?

Automation is the technology that uses automatic control system for operating equipment like machinery, processes in factories, etc. with minimal human assistance. It has three basic types - fixed automation, programmable automation, and flexible automation. This is an introductory chapter which will briefly introduce about automation, its types, advantages and disadvantages.

Chapter - Understanding Industrial Automation

Industrial automation refers to usage of the control systems, like computers and information technologies for handling and managing the different processes and machineries in an industry without human assistance. This chapter closely examines industrial automation to provide an extensive understanding of the subject.

Chapter - Control Systems in Automation

Control systems deal with the operation of control systems and instrumentation used in different processes related to automation. It includes discrete control, PID controller, sequential control, motion control, distributed control system, etc. The topics elaborated in this chapter will help in gaining a better perspective of these control systems in automation.

Chapter – Automation and Manufacturing

Automated manufacturing systems perform certain operations like processing, inspection, assembly, and material handling. It also includes computer-integrated manufacturing, automatic process control, automated production and assembly line, etc. This chapter has been carefully written to provide an easy understanding of the aspects related to automation and manufacturing.

Chapter – Industrial Automation Protocols

There are various industrial automation protocols that are used for building automation, power-system automation, vehicular automation, etc. Some of them are DC-BUS, DeviceNet, DNP3, EtherCAT, Modbus, MTConnect, OPC, etc. This chapter discusses the different industrial automation protocols in detail.

Geoffrey Williamson

1
What is Automation?

Automation is the technology that uses automatic control system for operating equipment like machinery, processes in factories, etc. with minimal human assistance. It has three basic types – fixed automation, programmable automation, and flexible automation. This is an introductory chapter which will briefly introduce about automation, its types, advantages and disadvantages.

Automation is the application of machines to tasks once performed by human beings or, increasingly, to tasks that would otherwise be impossible. Although the term mechanization is often used to refer to the simple replacement of human labour by machines, automation generally implies the integration of machines into a self-governing system. Automation has revolutionized those areas in which it has been introduced, and there is scarcely an aspect of modern life that has been unaffected by it. The term automation was coined in the automobile industry about 1946 to describe the increased use of automatic devices and controls in mechanized production lines. The origin of the word is attributed to D.S. Harder, an engineering manager at the Ford Motor Company at the time. The term is used widely in a manufacturing context, but it is also applied outside manufacturing in connection with a variety of systems in which there is a significant substitution of mechanical, electrical, or computerized action for human effort and intelligence.

In general usage, automation can be defined as a technology concerned with performing a process by means of programmed commands combined with automatic feedback control to ensure proper execution of the instructions. The resulting system is capable of operating without human intervention. The development of this technology has become increasingly dependent on the use of computers and computer-related technologies. Consequently, automated systems have become increasingly sophisticated and complex. Advanced systems represent a level of capability and performance that surpass in many ways the abilities of humans to accomplish the same activities.

Automation technology has matured to a point where a number of other technologies have developed from it and have achieved a recognition and status of their own. Robotics is one of these technologies; it is a specialized branch of automation in which the automated machine possesses certain anthropomorphic, or humanlike, characteristics. The most typical humanlike characteristic of a modern industrial robot is its powered mechanical arm. The robot's arm can be programmed to move through a sequence of motions to perform useful tasks, such as loading and unloading parts at a production machine or making a sequence of spot-welds on the sheet-metal parts of an automobile body during assembly. As these examples suggest, industrial robots are typically used to replace human workers in factory operations.

Principles and Theory of Automation

The three basic building blocks of automation are: (1) a source of power to perform some action, (2) feedback controls, and (3) machine programming. Almost without exception, an automated system will exhibit all these elements.

Power Source

An automated system is designed to accomplish some useful action, and that action requires power. There are many sources of power available, but the most commonly used power in today's automated systems is electricity. Electrical power is the most versatile, because it can be readily generated from other sources (e.g., fossil fuel, hydroelectric, solar, and nuclear) and it can be readily converted into other types of power (e.g., mechanical, hydraulic, and pneumatic) to perform useful work. In addition, electrical energy can be stored in high-performance, long-life batteries.

The actions performed by automated systems are generally of two types: (1) processing and (2) transfer and positioning. In the first case, energy is applied to accomplish some processing operation on some entity. The process may involve the shaping of metal, the molding of plastic, the switching of electrical signals in a communication system, or the processing of data in a computerized information system. All these actions entail the use of energy to transform the entity (e.g., the metal, plastic, electrical signals, or data) from one state or condition into another more valuable state or condition. The second type of action—transfer and positioning—is most readily seen in automated manufacturing systems designed to perform work on a product. In these cases, the product must generally be moved (transferred) from one location to another during the series of processing steps. At each processing location, accurate positioning of the product is generally required. In automated communications and information systems, the terms transfer and positioning refer to the movement of data (or electrical signals) among various processing units and the delivery of information to output terminals (printers, video display units, etc.) for interpretation and use by humans.

Feedback Controls

Feedback controls are widely used in modern automated systems. A feedback control system consists of five basic components: (1) input, (2) process being controlled, (3) output, (4) sensing elements, and (5) controller and actuating devices. The term closed-loop feedback control is often used to describe this kind of system.

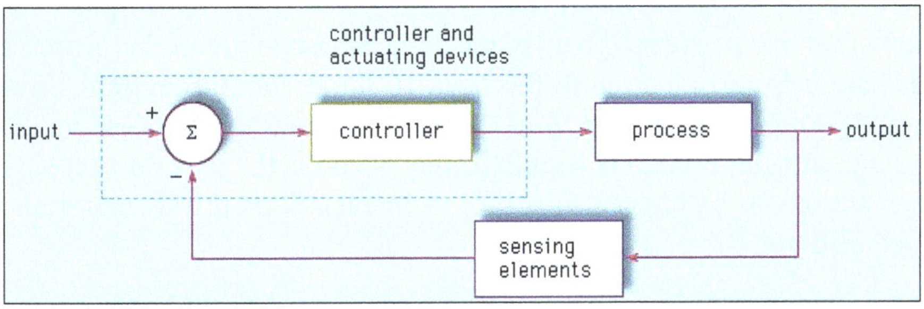

The components of a feedback control system and their relationships.

The input to the system is the reference value, or set point, for the system output. This represents the desired operating value of the output. Using the previous example of the heating system as an illustration, the input is the desired temperature setting for a room. The process being controlled is the heater (e.g., furnace). In other feedback systems, the process might be a manufacturing operation, the rocket engines on a space shuttle, the automobile engine in cruise control, or any of a variety of other processes to which power is applied. The output is the variable of the process that is being measured and compared to the input; in the above example, it is room temperature.

The sensing elements are the measuring devices used in the feedback loop to monitor the value of the output variable. In the heating system example, this function is normally accomplished using a bimetallic strip. This device consists of two metal strips joined along their lengths. The two metals possess different thermal expansion coefficients; thus, when the temperature of the strip is raised, it flexes in direct proportion to the temperature change. As such, the bimetallic strip is capable of measuring temperature. There are many different kinds of sensors used in feedback control systems for automation.

The purpose of the controller and actuating devices in the feedback system is to compare the measured output value with the reference input value and to reduce the difference between them. In general, the controller and actuator of the system are the mechanisms by which changes in the process are accomplished to influence the output variable. These mechanisms are usually designed specifically for the system and consist of devices such as motors, valves, solenoid switches, piston cylinders, gears, power screws, pulley systems, chain drives, and other mechanical and electrical components. The switch connected to the bimetallic strip of the thermostat is the controller and actuating device for the heating system. When the output (room temperature) is below the set point, the switch turns on the heater. When the temperature exceeds the set point, the heat is turned off.

Machine Programming

The programmed instructions determine the set of actions that is to be accomplished automatically by the system. The program specifies what the automated system should do and how its various components must function in order to accomplish the desired result. The content of the program varies considerably from one system to the next. In relatively simple systems, the program consists of a limited number of well-defined actions that are performed continuously and repeatedly in the proper sequence with no deviation from one cycle to the next. In more complex systems, the number of commands could be quite large, and the level of detail in each command could be significantly greater. In relatively sophisticated systems, the program provides for the sequence of actions to be altered in response to variations in raw materials or other operating conditions.

Programming commands are related to feedback control in an automated system in that the program establishes the sequence of values for the inputs (set points) of the various feedback control loops that make up the automated system. A given programming command may specify the set point for the feedback loop, which in turn controls some action that the system is to accomplish. In effect, the purpose of the feedback loop is to verify that the programmed step has been carried out. For example, in a robot controller, the program might specify that the arm is to move to a designated position, and the feedback control system is used to verify that the move has been correctly made.

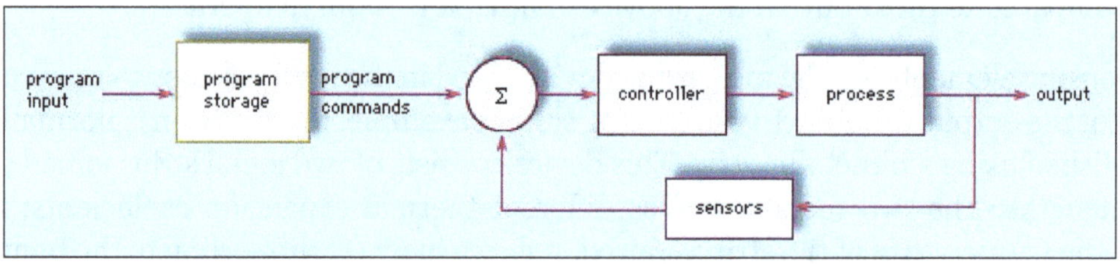

Relationship of program control and feedback control in an automated system.

Some of the programmed commands may be executed in a simple open-loop fashion—i.e., without the need for a feedback loop to verify that the command has been properly carried out. For example, a command to flip an electrical switch may not require feedback. The need for feedback control in an automated system might arise when there are variations in the raw materials being fed into a production process, and the system must take these variations into consideration by making adjustments in its controlled actions. Without feedback, the system would be unable to exert sufficient control over the quality of the process output.

The programmed commands may be contained on mechanical devices (e.g., mechanical cams and linkages), punched paper tape, magnetic tape, magnetic disks, computer memory, or any of a variety of other media that have been developed over the years for particular applications. It is common today for automated equipment to use computer storage technology as the means for storing the programmed commands and converting them into controlled actions. One of the advantages of computer storage is that the program can be readily changed or improved. Altering a program that is contained on mechanical cams involves considerable work.

Programmable machines are often capable of making decisions during their operation. The decision-making capacity is contained in the control program in the form of logical instructions that govern the operation of such a system under varying circumstances. Under one set of circumstances, the system responds one way; under different circumstances, it responds in another way. There are several reasons for providing an automated system with decision-making capability, including (1) error detection and recovery, (2) safety monitoring, (3) interaction with humans, and (4) process optimization.

Error detection and recovery is concerned with decisions that must be made by the system in response to undesirable operating conditions. In the operation of any automated system, malfunctions and errors sometimes occur during the normal cycle of operations, for which some form of corrective action must be taken to restore the system. The usual response to a system malfunction has been to call for human assistance. There is a growing trend in automation and robotics to enable the system itself to sense these malfunctions and to correct for them in some manner without human intervention. This sensing and correction is referred to as error detection and recovery, and it requires that a decision-making capability be programmed into the system.

Safety monitoring is a special case of error detection and recovery in which the malfunction involves a safety hazard. Decisions are required when the automated system sensors detect that a safety condition has developed that would be hazardous to the equipment or humans in the vicinity of the equipment. The purpose of the safety-monitoring system is to detect the hazard and to take the most appropriate action to remove or reduce it. This may involve stopping the operation and alerting maintenance personnel to the condition, or it may involve a more complex set of actions to eliminate the safety problem.

Automated systems are usually required to interact with humans in some way. An automatic bank teller machine, for example, must receive instructions from customers and act accordingly. In some automated systems, a variety of different instructions from humans is possible, and the decision-making capability of the system must be quite sophisticated in order to deal with the array of possibilities.

A fourth reason for decision making in an automated system is to optimize the process. The need for optimization occurs most commonly in processes in which there is an economic performance criterion whose optimization is desirable. For example, minimizing cost is usually an important objective in manufacturing. The automated system might use adaptive control to receive appropriate sensor signals and other inputs and make decisions to drive the process toward the optimal state.

Industrial Robotics

Industrial robotics is an automation technology that has received considerable attention since about 1960.

The Robot Manipulator

The most widely accepted definition of an industrial robot is one developed by the Robotic Industries Association:

> "An industrial robot is a reprogrammable, multifunctional manipulator designed to move materials, parts, tools, or specialized devices through variable programmed motions for the performance of a variety of tasks."

The technology of robotics is concerned with the design of the mechanical manipulator and the computer systems used to control it. It is also concerned with the industrial applications of robots.

The mechanical manipulator of an industrial robot is made up of a sequence of link and joint combinations. The links are the rigid members connecting the joints. The joints (also called axes) are the movable components of the robot that cause relative motion between adjacent links. As shown in figure, there are five principal types of mechanical joints used to construct the manipulator. Two of the joints are linear, in which the relative motion between adjacent links is translational, and three are rotary types, in which the relative motion involves rotation between links.

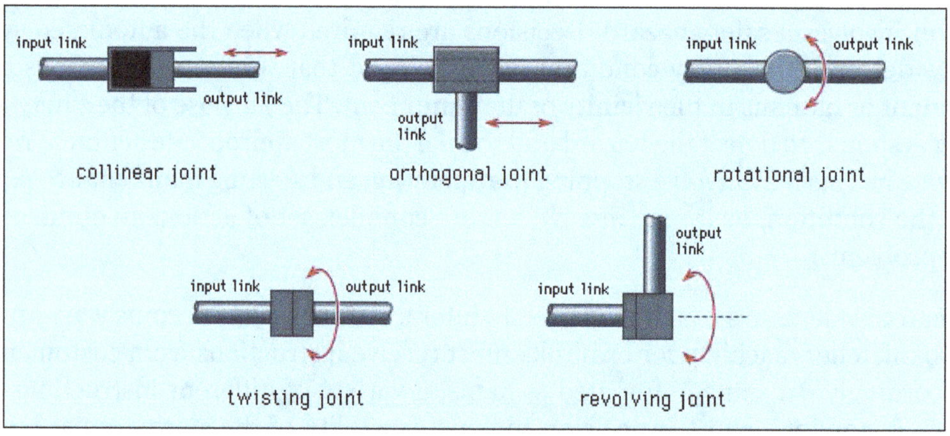

Five types of mechanical joints used in robot manipulators. Collinear and orthogonal are translational joints; rotational, twisting, and revolving are rotary joints.

The manipulator can be divided into two sections: (1) an arm-and-body, which usually consists of three joints connected by large links, and (2) a wrist, consisting of two or three compact joints. Attached to the wrist is a gripper to grasp a work part or a tool (e.g., a spot-welding gun) to perform a process. The two manipulator sections have different functions: the arm-and-body is used to move and position parts or tools in the robot's work space, while the wrist is used to orient the parts or tools at the work location. The arm-and-body section of most commercial robots is based on one of four configurations. Each of the anatomies, as they are sometimes called, provides a different work envelope (i.e., the space that can be reached by the robot's arm) and is suited to different types of applications.

Robot Programming

The computer system that controls the manipulator must be programmed to teach the robot the particular motion sequence and other actions that must be performed in order to accomplish its task. There are several ways that industrial robots are programmed. One method is called lead-through programming. This requires that the manipulator be driven through the various motions needed to perform a given task, recording the motions into the robot's computer memory. This can be done either by physically moving the manipulator through the motion sequence or by using a control box to drive the manipulator through the sequence.

A second method of programming involves the use of a programming language very much like a computer programming language. However, in addition to many of the capabilities of a computer programming language (i.e., data processing, computations, communicating with other computer devices, and decision making), the robot language also includes statements specifically designed for robot control. These capabilities include (1) motion control and (2) input/output. Motion-control commands are used to direct the robot to move its manipulator to some defined position in space. For example, the statement "move P1" might be used to direct the robot to a point in space called P1. Input/output commands are employed to control the receipt of signals from sensors and other devices in the work cell and to initiate control signals to other pieces of equipment in the cell. For instance, the statement "signal 3, on" might be used to turn on a motor in the cell, where the motor is connected to output line 3 in the robot's controller.

Manufacturing Applications of Automation and Robotics

One of the most important application areas for automation technology is manufacturing. To many people, automation means manufacturing automation. Three types of automation in production can be distinguished: (1) fixed automation, (2) programmable automation, and (3) flexible automation.

Fixed automation, also known as "hard automation," refers to an automated production facility in which the sequence of processing operations is fixed by the equipment configuration. In effect, the programmed commands are contained in the machines in the form of cams, gears, wiring, and other hardware that is not easily changed over from one product style to another. This form of automation is characterized by high initial investment and high production rates. It is therefore suitable for products that are made in large volumes. Examples of fixed automation include machining transfer lines found in the automotive industry, automatic assembly machines, and certain chemical processes.

Programmable automation is a form of automation for producing products in batches. The products are made in batch quantities ranging from several dozen to several thousand units at a time. For each new batch, the production equipment must be reprogrammed and changed over to accommodate the new product style. This reprogramming and changeover take time to accomplish, and there is a period of nonproductive time followed by a production run for each new batch. Production rates in programmable automation are generally lower than in fixed automation, because the equipment is designed to facilitate product changeover rather than for product specialization. A numerical-control machine tool is a good example of programmable automation. The program is coded in computer memory for each different product style, and the machine tool is controlled by the computer program. Industrial robots are another example.

Flexible automation is an extension of programmable automation. The disadvantage with programmable automation is the time required to reprogram and change over the production equipment for each batch of new product. This is lost production time, which is expensive. In flexible automation, the variety of products is sufficiently limited so that the changeover of the equipment can be done very quickly and automatically. The reprogramming of the equipment in flexible automation is done off-line; that is, the programming is accomplished at a computer terminal without using the production equipment itself. Accordingly, there is no need to group identical products into batches; instead, a mixture of different products can be produced one right after another.

Automated Production Lines

An automated production line consists of a series of workstations connected by a transfer system to move parts between the stations. This is an example of fixed automation, since these lines are typically set up for long production runs, perhaps making millions of product units and running for several years between changeovers. Each station is designed to perform a specific processing operation, so that the part or product is constructed stepwise as it progresses along the line. A raw work part enters at one end of the line, proceeds through each workstation, and emerges at the other end as a completed product. In the normal operation of the line, there is a work part being processed at each station, so that many parts are being processed simultaneously and a finished part is produced with each cycle of the line. The various operations, part transfers, and other activities taking place on an automated transfer line must all be sequenced and coordinated properly for the line to operate efficiently. Modern automated lines are controlled by programmable logic controllers, which are special computers that facilitate connections with industrial equipment (such as automated production lines) and can perform the kinds of timing and sequencing functions required to operate such equipment.

Automated production lines are utilized in many industries, most notably automotive, where they are used for processes such as machining and pressworking. Machining is a manufacturing process in which metal is removed by a cutting or shaping tool, so that the remaining work part is the desired shape. Machinery and motor components are usually made by this process. In many cases, multiple operations are required to completely shape the part. If the part is mass-produced, an automated transfer line is often the most economical method of production. The many separate operations are divided among the workstations. Transfer lines date back to about 1924.

Pressworking operations involve the cutting and forming of parts from sheet metal. Examples of such parts include automobile body panels, outer shells of major appliances (e.g., laundry machines and ranges), and metal furniture (e.g., desks and file cabinets). More than one processing step is often required to complete a complicated part. Several presses are connected together in sequence by handling mechanisms that transfer the partially completed parts from one press to the next, thus creating an automated pressworking line.

Automated Assembly

Assembly operations have traditionally been performed manually, either at single assembly workstations or on assembly lines with multiple stations. Owing to the high labour content and high cost of manual labour, greater attention has been given in recent years to the use of automation for assembly work. Assembly operations can be automated using production line principles if the quantities are large, the product is small, and the design is simple (e.g., mechanical pencils, pens, and cigarette lighters). For products that do not satisfy these conditions, manual assembly is generally required.

Automated assembly machines have been developed that operate in a manner similar to machining transfer lines, with the difference being that assembly operations, instead of machining, are performed at the workstations. A typical assembly machine consists of several stations, each equipped with a supply of components and a mechanism for delivering the components into position for assembly. A workhead at each station performs the actual attachment of the component. Typical workheads include automatic screwdrivers, staking or riveting machines, welding heads, and other joining devices. A new component is added to the partially completed product at each workstation, thus building up the product gradually as it proceeds through the line. Assembly machines of this type are considered to be examples of fixed automation, because they are generally configured for a particular product made in high volume. Programmable assembly machines are represented by the component-insertion machines employed in the electronics industry.

Flexible Manufacturing Systems

A flexible manufacturing system (FMS) is a form of flexible automation in which several machine tools are linked together by a material-handling system, and all aspects of the system are controlled by a central computer. An FMS is distinguished from an automated production line by its ability to process more than one product style simultaneously. At any moment, each machine in the system may be processing a different part type. An FMS can also cope with changes in product mix and production schedule as demand patterns for the different products made on the system change over time. New product styles can be introduced into production with an FMS, so long as they fall within the range of products that the system is designed to process. This kind of system is therefore ideal when demand for the products is low to medium and there are likely to be changes in demand.

The components of an FMS are (1) processing machines, which are usually CNC machine tools that perform machining operations, although other types of automated workstations such as inspection stations are also possible, (2) a material-handling system, such as a conveyor system, which is capable of delivering work parts to any machine in the FMS, and (3) a central computer system that is responsible for communicating NC part programs to each machine and for coordinating the activities of the machines and the material-handling system. In addition, a fourth component of an FMS is human labour. Although the flexible manufacturing system represents a high level of production automation, people are still needed to manage the system, load and unload parts, change tools, and maintain and repair the equipment.

Computer-integrated Manufacturing

Since about 1970 there has been a growing trend in manufacturing firms toward the use of computers to perform many of the functions related to design and production. The technology associated with this trend is called CAD/CAM, for computer-aided design and computer-aided manufacturing. Today it is widely recognized that the scope of computer applications must extend beyond design and production to include the business functions of the firm. The name given to this more comprehensive use of computers is computer-integrated manufacturing (CIM).

CAD/CAM is based on the capability of a computer system to process, store, and display large amounts of data representing part and product specifications. For mechanical products, the data represent graphic models of the components; for electrical products, they represent circuit information; and so forth. CAD/CAM technology has been applied in many industries, including machined components, electronics products, and equipment design and fabrication for chemical processing. CAD/CAM involves not only the automation of the manufacturing operations but also the automation of elements in the entire design-and-manufacturing procedure.

Computer-aided design (CAD) makes use of computer systems to assist in the creation, modification, analysis, and optimization of a design. The designer, working with the CAD system rather than the traditional drafting board, creates the lines and surfaces that form the object (product, part, structure, etc.) and stores this model in the computer database. By invoking the appropriate CAD software, the designer can perform various analyses on the object, such as heat transfer calculations. The final object design is developed as adjustments are made on the basis of these analyses. Once the design procedure has been completed, the computer-aided design system can generate the detailed drawings required to make the object.

Computer-aided manufacturing (CAM) involves the use of computer systems to assist in the planning, control, and management of production operations. This is accomplished by either direct or indirect connections between the computer and production operations. In the case of the direct connection, the computer is used to monitor or control the processes in the factory. Computer process monitoring involves the collection of data from the factory, the analysis of the data, and the communication of process-performance results to plant management. These measures increase the efficiency of plant operations. Computer process control entails the use of the computer system to execute control actions to operate the plant automatically, as described above. Indirect connections between the computer system and the process involve applications in which the computer supports the production operations without actually monitoring or controlling them. These applications include planning and management functions that can be performed by the computer (or by humans working with the computer) more efficiently than by humans alone. Examples of these functions are planning the step-by-step processes for the product, part programming in numerical control, and scheduling the production operations in the factory.

Computer-integrated manufacturing includes all the engineering functions of CAD/CAM and the business functions of the firm as well. These business functions include order entry, cost accounting, employee time records and payroll, and customer billing. In an ideal CIM system, computer technology is applied to all the operational and information-processing functions of the company, from customer orders through design and production (CAD/CAM) to product shipment and customer service. The scope of the computer system includes all activities that are concerned with manufacturing. In many ways, CIM represents the highest level of automation in manufacturing.

Automation in Daily Life

In addition to the manufacturing applications of automation technology, there have been significant achievements in such areas as communications, transportation, service industries, and consumer products.

Communications

One of the earliest practical applications of automation was in telephone switching. The first switching machines, invented near the end of the 19th century, were simple mechanical switches that were remotely controlled by the telephone user pushing buttons or turning a dial on the phone. Modern electronic telephone switching systems are based on highly sophisticated digital computers that perform functions such as monitoring thousands of telephone lines, determining which lines require service, storing the digits of each telephone number as it is being dialed, setting up the required connections, sending electrical signals to ring the receiver's phone, monitoring the call during its progress, and disconnecting the phone when the call is completed. These systems also are used to time and bill toll calls and to transmit billing information and other data relative to the business operations of the phone company. The newest electronic systems automatically transfer calls to alternate numbers, call back the user when busy lines become free, and perform other customer services in response to dialed codes. These systems also perform function tests on their own operations, diagnose problems when they arise, and print out detailed instructions for repairs.

Other applications of automation in communications systems include local area networks, communications satellites, and automated mail-sorting machines. A local area network (LAN) operates like an automated telephone company within a single building or group of buildings. Local area networks are generally capable of transmitting not only voice but also digital data between terminals in the system. Communications satellites have become essential for communicating telephone or video signals across great distances. Such communications would not be possible without the automated guidance systems that place and retain the satellites in predetermined orbits. Automatic mail-sorting machines have been developed for use in many post offices throughout the world to read codes on envelopes and sort the envelopes according to destination.

Transportation

Automation has been applied in various ways in the transportation industries. Applications include airline reservation systems, automatic pilots in aircraft and locomotives, and urban mass-transit systems. The airlines use computerized reservation systems to continuously monitor the status of all flights. With these systems, ticket agents at widely dispersed locations can obtain information about the availability of seats on any flight in a matter of seconds. The reservation systems compare requests for space with the status of each flight, grant space when available, and automatically update the reservation status files. Passengers can even receive their seat assignments well in advance of flight departures.

Nearly all commercial aircraft are equipped with instruments called automatic pilots. Under normal flying conditions, these systems guide an airplane over a predetermined route by detecting changes in the aircraft's orientation and heading from gyroscopes and similar instruments and by providing appropriate control signals to the plane's steering mechanism. Automatic navigation systems and instrument landing systems operate by using radio signals from ground beacons that provide the aircraft with course directions for guidance. When an airplane is within the traffic pattern for ground control, its human pilot normally assumes control.

Examples of automated rail transportation include American urban mass-transit systems such as BART (Bay Area Rapid Transit) in San Francisco; MARTA (Metropolitan Atlanta Rapid Transit Authority) in Atlanta, Ga.; and the Metrorail in Washington, D.C.

The BART system serves as a useful example; it consists of more than 75 miles (120 kilometres) of track, with about 100 trains operating at peak hours between roughly 30 stations. The trains sometimes attain speeds of 80 miles per hour with intervals between trains of as little as 90 seconds. In each train there is one operator whose role is that of an observer and communicator and who can override the automatic system in case of emergency. The automatic system protects the trains by assuring a safe distance between them and by controlling their speed. Another function of the system is to control train routings and make adjustments in the operation of each train to keep the entire system operating on schedule.

As a train enters the station, it automatically transmits its identification, destination, and length, thus lighting up a display board for passenger information and transmitting information to the control centres. Signals are automatically returned to the train to regulate its time in the station and its running time to the next station. At the beginning of the day, an ideal schedule is determined; as the day progresses, the performance of each train is compared with the schedule, and adjustments are made to each train's operation as required. The entire system is controlled by two identical computers, so that if one malfunctions, the other assumes complete control. In the event of a complete failure of the computer control system, the system reverts to manual control.

Service Industries

Automation of service industries includes an assortment of applications as diverse as the services themselves, which include health care, banking and other financial services, government, and retail trade.

In health care the use of automation in the form of computer systems has increased dramatically to improve services and relieve the burden on medical staffs. In hospitals computer terminals on each nursing care floor record data on patient status, medications administered, and other relevant information. Some of these systems are used to perform additional functions such as ordering drugs from the hospital pharmacy and calling for orderlies. The system provides an official record of the nursing care given to patients and is used by the nursing staff to give a report at shift-change time. The computer system is connected to the hospital's business office so that proper charges can be made to each patient's account for services rendered and medicines provided.

Robotics is likely to play a role in future health care delivery systems. The work that is done in hospitals by nurses, orderlies, and similar staff personnel includes some tasks that are routine and repetitive. Duties that might be automated using robots include making beds, delivering linens, and moving supplies between locations in the hospital. Robots might even become involved in certain aspects of patient care such as transporting patients to services in the hospital, passing food trays, and similar functions in which it is not critical that a hospital staff member be present. Research is currently under way to develop robots that would be capable of providing assistance to paraplegics and other physically handicapped persons. These robots would respond to voice commands and would be able to interpret statements in natural language (e.g., everyday English) from patients requesting service.

Banking and financial institutions have embraced automation in their operations—principally through computer technology—to facilitate the processing of large volumes of documents and financial transactions. The sorting of checks is done by optical character-recognition systems utilizing the special alphanumeric characters at the bottom of checks. Bank balances are computed and recorded using computer systems installed by virtually all financial institutions. Major banks have established electronic banking systems, including automatic teller machines. Located in places convenient for their customers, these automatic tellers permit users to complete basic transactions without requiring the assistance of bank personnel.

The stock exchanges rely on computer-automated systems to report transactions by ticker tape or closed circuit television. Brokerage houses use a computerized record-keeping system to track their customers' accounts. Monthly statements indicating the status of each account are automatically prepared and mailed to customers. Account executives employ video monitors in their offices, backed by a massive database, to retrieve current information on each stock almost instantaneously while they discuss possible purchases with their clients. Stock certificates are typically issued with machine-readable identifications to facilitate record keeping in sales and exchanges.

Credit card transactions have also become highly automated. Restaurants, retailers, and other organizations are using systems that automatically check the validity of a credit card and the credit standing of the cardholder in a matter of seconds as the customer waits for the transaction to be finalized. Some credit card transactions trigger immediate transfer of funds equal to the amount of the sale from the cardholder's account into the merchant's account.

Many government services are automated by means of computers and computerized databases. The Internal Revenue Service (IRS) of the U.S. government must review and approve the tax returns of millions of taxpayers each year. The detailed checking of returns is a task that has traditionally been done by large staffs of professional auditors on a sampled basis. In 1985 the IRS began using a computerized system to automate the auditing procedure for the 1984 returns. This system is programmed to perform the complex tax calculations on each return being audited. As tax laws change, the system is reprogrammed to do the calculations for the year. The computerized auditing system has permitted a substantial increase in the work load of the IRS auditing department without a corresponding increase in staffing.

Retail trade has seen a number of changes in its operations as a result of automation. Selling merchandise has typically been a very labour-intensive activity, with sales associates needed to assist customers with their selections and then finalize transactions at the cash register. Each transaction depletes the store's inventory, so the item purchased must be identified for reorder. Much clerical effort is expended by the store when inventory is managed by strictly manual procedures. Computerized systems have been installed in most modern retail stores to speed sales transactions and automatically update inventory records as the stock of each item is reduced. The systems are based on the Universal Product Code (UPC), originally adopted by the grocery industry in 1973, which uses optical bar-code technology. A bar code is an identification symbol consisting of a series of wide and narrow bars attached to each product that can be scanned and recognized by a bar-code reader. At the cash registers, these readers quickly identify the items being purchased. As the sales associate scans across the symbol using a laser beam reader, the product is properly identified and its price is entered into the sales transaction. Simultaneously, a record of the sale is made in the inventory files so that the item can be reordered.

Consumer Products

Consumer products ranging from automobiles to small appliances have been automated for the benefit of the user. Microwave ovens, washing machines, dryers, refrigerators, video recorders, and other modern household appliances typically contain a microprocessor that works as the computer controller for the device. The consumer operates the appliance by programming the controller to perform the required functions, including timing (ovens, dryers), power levels (microwave ovens), input channels (video recorders), and other cycle options (washing machines). The programming of the device is done simply by pressing a series of buttons in the proper sequence, so the user does not think of the procedure as programming a computer.

The automobile is an example of a highly automated consumer product. The modern automobile is typically equipped with several microprocessors that operate a variety of functions, including engine control (fuel-air ratio, for example), the clock, the radio, and cruise control.

Automation and Society

Over the years, the social merits of automation have been argued by labour leaders, business executives, government officials, and college professors. The biggest controversy has focused on how automation affects employment. There are other important aspects of automation, including its effect on productivity, economic competition, education, and quality of life.

Impact on the Individual

Nearly all industrial installations of automation, and in particular robotics, involve a replacement of human labour by an automated system. Therefore, one of the direct effects of automation in factory operations is the dislocation of human labour from the workplace. The long-term effects of automation on employment and unemployment rates are debatable. Most studies in this area have been controversial and inconclusive. Workers have indeed lost jobs through automation, but population increases and consumer demand for the products of automation have compensated for these losses. Labour unions have argued, and many companies have adopted the policy, that workers displaced by automation should be retrained for other positions, perhaps increasing their skill levels in the process. This argument succeeds so long as the company and the economy in general are growing at a rate fast enough to create new positions as the jobs replaced by automation are lost.

Of particular concern for many labour specialists is the impact of industrial robots on the work force, since robot installations involve a direct substitution of machines for humans, sometimes at a ratio of two to three humans per robot. The opposing argument within the United States is that robots can increase productivity in American factories, thereby making these firms more competitive and ensuring that jobs are not lost to overseas companies. The effect of robotics on labour has been relatively minor, because the number of robots in the United States is small compared with the number of human workers. As of the early 1990s, there were fewer than 100,000 robots installed in American factories, compared with a total work force of more than 100 million persons, about 20 million of whom work in factories.

Automation affects not only the number of workers in factories but also the type of work that is done. The automated factory is oriented toward the use of computer systems and sophisticated programmable machines rather than manual labour. Greater emphasis is placed on knowledge-based work and technical skill rather than physical work. The types of jobs found in modern factories include more machine maintenance, improved scheduling and process optimization, systems analysis, and computer programming and operation. Consequently, workers in automated facilities must be technologically proficient to perform these jobs. Professional and semiprofessional positions, as well as traditional labour jobs, are affected by this shift in emphasis toward factory automation.

Impact on Society

Besides affecting individual workers, automation has an impact on society in general. Productivity is a fundamental economic issue that is influenced by automation. The productivity of a process is traditionally defined as the ratio of output units to the units of labour input. A properly justified automation project will increase productivity owing to increases in production rate and reductions in labour content. Over the years, productivity gains have led to reduced prices for products and increased prosperity for society.

A number of issues related to education and training have been raised by the increased use of automation, robotics, computer systems, and related technologies. As automation has increased, there has developed a shortage of technically trained personnel to implement these technologies competently. This shortage has had a direct influence on the rate at which automated systems can be introduced. The shortage of skilled staffing in automation technologies raises the need for vocational and technical training to develop the required work-force skills. Unfortunately the educational system is also in need of technically qualified instructors to teach these subjects, and the laboratory equipment available in schools does not always represent the state-of-the-art technology typically used in industry.

Types of Automation

The automation system replaces the conventional hard-wired relay system by automating the process functions with the use of various hardware and software tools with minimal human intervention. There are different types of automation systems are used in different applications, but all automated systems are not similar. Depends on application area and control nature, these automated systems are classified. Some of these automation types are:

Industrial Automation

Industrial automation is the process of making industrial production processes more flexible and simpler at higher efficiencies. Automation integration to the industries results intelligent manufacturing solutions at improved product quality and productivity with reduced downtime and wastes. Industrial automation involves in usage of usage of various control devices such as PC/PLC's, various sensors and actuators, communication buses/modules, machine drives, HMI (Human Machine Interface) systems and other control equipments.

This type automation is most popularly used in automotive, computer and electronics, medical, telecommunication, consumer goods and other industrial applications. The automation systems can be a fixed, programmed, flexible and integrated systems.

Numerically Controlled Machines

These machines are of computer controlled machines which uses computers to perform the control operations by acquiring, processing, calculating and controlling the process variables. This automation is a programmed version of machine tools and also called as Computerized Numerical Controlled (CNC) Machines. These CNC machines are used in cutting and milling applications for high accuracy and accurate precision operation.

Computer – Aided Manufacturing (CAM)

In this, the entire manufacturing process (includes production, planning and control) is automated with the use of numerically controlled machines, industrial robots and other types of automation devices. These automation systems also make use of computers to plan, design and layout the various products. Examples of this automation systems are computer-aided design (CAD), computer-aided design and drafting (CADD) and computer-aided process planning (CAPP).

Industrial Robots

Industrial Robots.

These are a type of automated machines or equipment's that can perform the different tasks for longer duration. These are mostly implemented in the areas that are highly dangerous or hazardous for humans.

Flexible Manufacturing Systems

This automation is of fully automated one. Starting from planning and designing process to dispatching of products, the whole system is completely integrated to be automated. This automation combines numerically controlled machines, industrial robots and other automation equipment's into one integrated system.

Home Automation

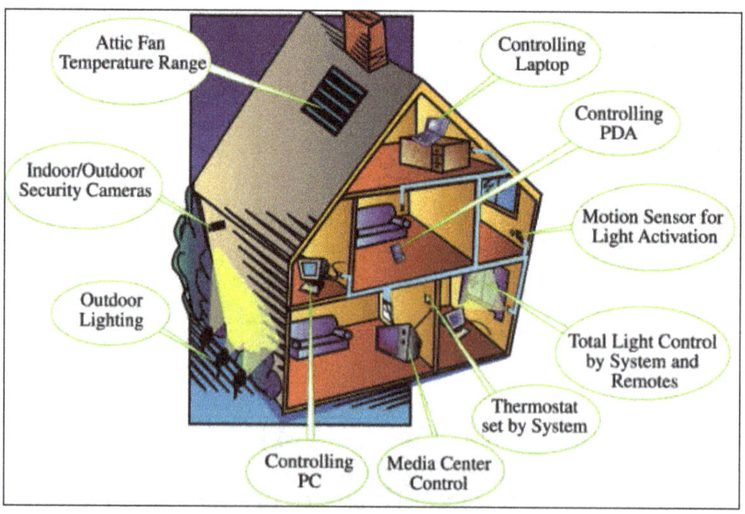

Home Automation.

As the technologies are rapidly advancing that affects the human daily lives more flexible, particularly in controlling home appliances. Electrical installations are the heart of every building, therefore intelligent building control absolutely ensures the safety and efficient control and eventually saves the electric power consumption and human energy. Home automation system does the operations like lighting control and regulation, regulating HVAC (Heating, Ventilation and Air Conditioning) equipment's, energy and load management, security and monitoring, audio/video systems, HMI's and other control tasks.

Home automation uses the various sensors to sense parameters like temperature, pressure, motion, illumination, etc. and sends these signals to central controllers. These controllers are programmable devices like PLCs which are programmed according to the application it is employed for. Therefore, these controllers get the inputs from sensors and correspondingly sends the control signals to actuating devices like relays. Some of the types of home automation system are given below.

- Power Line Home Automation System: This automation is of low cost since it uses power lines to transfer the information or data therefore it doesn't require additional cables for transferring the information. However, this system is of little bit complex one that necessitates additional converter circuits.
- Wired Home Automation System: This type of automation uses a common communication bus to which all the equipments in homes are connected. This communication bus or cable is connected to a main controller (programmable logic controller) in order to acquire input signals and to send control command signals to the actuators.
- Wireless Home Automation: This is an advanced automation technology as compared with wired automation. This automaton uses wireless technologies like RF, Zigbee, GSM, WI-Fi and Bluetooth for achieving remote control operation. This is a flexible type of automation requires negligible wires to connect different appliances in homes to the central control unit.

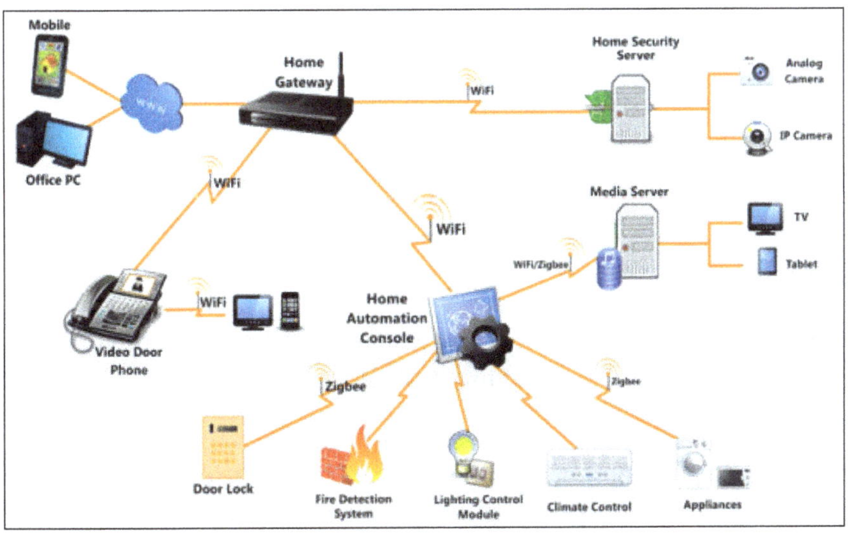

Wireless Home Automation.

- Wireless Home Automation: These are the two types of automation system that has been implementing in most of the industries and homes. Overall quality, accuracy, productivity, reduced labor costs and safety are better arguments for the automation systems.

Advantages of Automation

Given the right tools, automating computer operations can be surprisingly easy and can reap major benefits. Understanding these benefits—and some obstacles—will help you develop support for an operations automation project. A recent study by a leading trade journal asked the question, "What do you see as the most important benefits of an automated or unattended computer center?" The primary benefits of operations automation cited most often were cost reduction, productivity, availability, reliability, and performance.

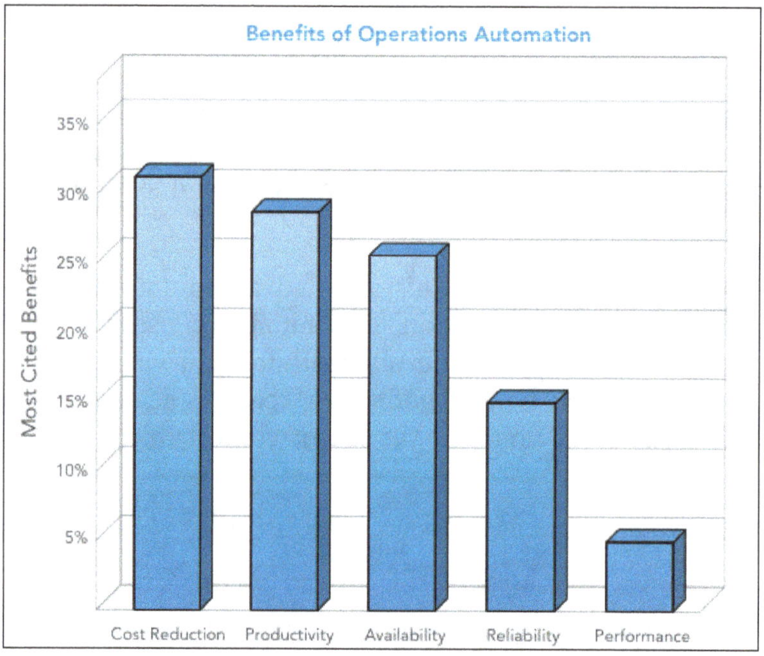

Principle Benefits of Automation.

Cost Reduction

Every business faces global pressure to increase their profitability. One approach is to reduce costs. But, reducing the capabilities of the computer center negatively impacts the entire company.

Automation software is a better and more intelligent approach to cost containment and reduction. The greatest opportunity is to increase service to the customer (end user) while systematically reducing costs. Management often overlooks this potential for savings. Most modern servers have a low operating cost and the total cost of ownership has been declining. Even so, the cost of the operations staff can be as high as 71% of the total cost.

Productivity

As an organization's technology demands grow, productivity becomes a bigger concern. Typically, as other business areas were given tools to increase their productivity and effectiveness, IT operations took a back seat. The proliferation of desktop productivity software has created substantial gains in the office and HR environments. But, instead of alleviating workload for the IT professionals in the back room, the spread of PCs has meant more tasks to be accomplished.

As people use computers more, they place greater demands on the system. More users are generating more jobs, and printed output has increased despite efforts to reduce printed reports. In spite of the trend to online transaction-oriented and client/server systems, batch workloads continue to grow. Production batch jobs still consume the majority of CPU time, and in large shops, jobs are constantly being added.

Automated operations can resolve these issues in several ways. Job scheduling software increases batch throughput by automating the production batch schedule. In the early days, computer throughput was limited by how fast operators could reset switches on the console. The solution for today is not to allow the computer to remain idle while waiting for the operator to release the next job. You save time and money by eliminating the lag time between jobs and minimizing operator intervention. In some instances, you can trim hours off your process, which lets you process more work and significantly improve system use.

Once the job schedule is established, the automation software executes the commands precisely and in the correct sequence, eliminating operator errors. Forecasting job completion and being able to perform "what if" analyses of schedule changes benefits operations by removing much of the guesswork from daily tasks.

Availability

Companies are continually more reliant on their computers. Day-to-day business is routinely conducted with online systems: order entry, reservations, assembly instructions, shipping orders—the list goes on. If the computer is not available, the business suffers.

Years ago, it was considered acceptable to have the computer unavailable for a few hours. Today, with the high volume of cloud computing, the outage of key systems can cost millions of dollars in lost revenue and tarnish a company's reputation.

High availability is clearly one of IT management's primary goals. Here too, automated operations can help. A disk drive may crash, but the situation becomes serious when there is not an adequate backup— or worse, the tape cannot be found. A key advantage to automation is the ability to automate your save and recovery systems to ensure protection from the potential disaster of disk loss, or inadvertent damage to system objects from human error.

In a networked environment, centralized management also makes sense. Remote resources can solve business issues while a single operator at a central console observes critical functions throughout the network. Continuous monitoring with a low CPU and communications overhead makes it easier to spot vital network performance trends.

Reliability

Productivity is an obvious benefit of automation. However, reliability is the real gem that sparkles with automation. It is the cornerstone of any good IT operations department and without it you have confusion, chaos, and unhappy users. IT operations requires two opposed skill sets: On one hand, an operations person needs highly technical skills, such as the ability to understand the complexities of an operating system and to analyze and solve problems as they arise. On the other hand, this same person has to be content pushing buttons and loading paper.

Automated operations ensure that jobs are not forgotten or run out of sequence, that prerequisite jobs are completed successfully, that the input data is correct, and that any special processing is performed.

All these types of errors occur in single-location organizations. Now, imagine a network of multiple systems, geographically dispersed that include multiple operating systems, communications issues, integrated local area network processing, and attached PCs. The chance for errors rises exponentially. The only way to make this type of environment work is automated operations.

Software can handle complex tasks dynamically and intelligently, based on predefined parameters. Yet, critical company functions such as releasing jobs, performing backups, and ensuring communications, are normally performed by entry-level individuals within the IT organization. The benefit of an automated system is that these functions are reliably executed by the automation software, relieving operations personnel from hours of tedious, boring and manual tasks.

Performance

Every company would like to have their enterprise perform like a thoroughbred. In reality, it is more likely to be overburdened with work. Even though advancements in computers make them faster and less expensive every year, the demands on them always catch up and eventually exceed the level of capability that a company's computer infrastructure possesses. That leaves a lot of companies wanting to improve their system performance.

Two options to improve performance are to upgrade hardware or purchase a newer system—both expensive choices. It's also possible to tune a system for better performance, but this takes a highly skilled person who is not normally available 24 hours a day. And, once a system is tuned for a specific workload, if the workload changes, the settings are no longer optimum.

Disadvantages of Automation

- Capital Expenditure: Whilst automation can prove highly effective and bring you a positive ROI, it may also require a fairly high capital cost. That's why, before making a decision we recommend considering both the investment needed and also the ROI you expect to achieve. When calculating the ROI it is important to include increased throughput value, reduced labour costs and the reduction in defects/recalls along with the capital expenditure before deciding whether or not there is a business case for investment.

- Gets rid of jobs: It is true that with the introduction of automation there are some jobs that may become redundant, but this does not necessarily have to be a negative implication of automation. Instead of staff performing mind-numbing, monotonous or unpleasant tasks, they can be trained to transfer to working in other areas of your business. Many companies have found that after installation of automation they have seen sales rise, thus creating more jobs in different parts of their business.

- Bespoke automation becomes redundant when production processes change: As with any type of machinery, if you change your production process or product you are manufacturing so that a particular machine is no longer part of the process then the machine becomes redundant. Therefore it is very important to future proof any automation you install into your production process. A skilled automation company will design your automation system to enable it to be easily adapted to suit changes in your product design or production process. For example; by using standard flexible automation such as robots, these can be easily used somewhere else in a manufacturing process even if the existing process becomes redundant.

References

- Automation, technology: britannica.com, Retrieved 27 January, 2020
- Different-types-of-automation-systems: elprocus.com, Retrieved 17 June, 2020
- Automated-operations-5-benefits-your-organization, resources, guides: helpsystems.com, Retrieved 15 April, 2020
- Advantages-and-disadvantages-of-automation: granta-automation.co.uk, Retrieved 23 March, 2020

2
Understanding Industrial Automation

Industrial automation refers to usage of the control systems, like computers and information technologies for handling and managing the different processes and machineries in an industry without human assistance. This chapter closely examines industrial automation to provide an extensive understanding of the subject.

Industrial automation is the use of various control devices like PC's/PLC's/DCS, used to have control on various operations of an industry without significant intervention from humans and to provide automatic control performance. In industries, control strategies use a set of technologies which are implemented to get the desired performance or output, making the automation system most essential for industries.

Automated process in industries.

Industrial automation involves usage of advanced control strategies like cascade controls, modern control hardware devices as PLC's, sensors and other instruments for sensing the control variables, signal conditioning equipments to connect the signals to the control devices, drives and other significant final control devices, standalone computing systems, communication systems, alarming and HMI (Human Machine Interface) systems.

Need of Automated Industry

- To reduce Periodic or manual checking: In some critical applications periodic checking of the process variable is necessary to perform industrial operations. Automation equipment reduces the periodic or manual operations and establishes the automatic working conditions.
- To increase the Productivity: Automating the manufacturing and other production processes increases the production rate by producing output at greater amounts for a given labour input.
- Reduce the Production Cost: Using the automatic machines and equipments, human intervention to control the processes abruptly falls. This reduces the investment on the labor cost hence the production cost.
- To improve Product Quality: Continuously doing the same work may not be perfect in all the cases in terms of quality specifications with human efforts. With automation equipment, one can get reliable and uniform product quality by using real time hardware control devices.
- To increase the Flexibility: Using the automation equipment various, process are handled simply without getting any complex environment particularly in manufacturing processes.
- Operator Friendly and Improves the Safety: Complexity of operating the equipments or processes is reduced with industrial automation. It changes the position of the operator as operator to the supervisory role.

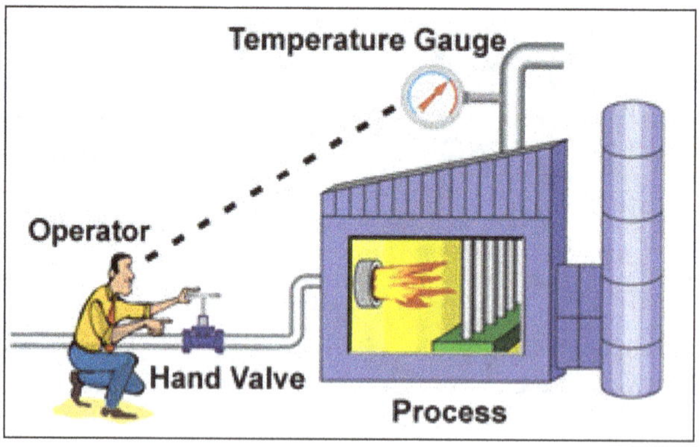

Manual control.

Structure of Industrial Automation

Structure of the industrial automation explains various levels of operation. These include sensor level, automation control level (Unit, cell, process controls), supervision level and enterprise level. Pyramid structure indicates that, as you go up the tip, the information is aggregated and while coming down it is dissolved. This means we will get the detailed information for a particular variable at the bottom. Industrial automation doesn't mean that all the levels are automated like enterprise level need not be automated.

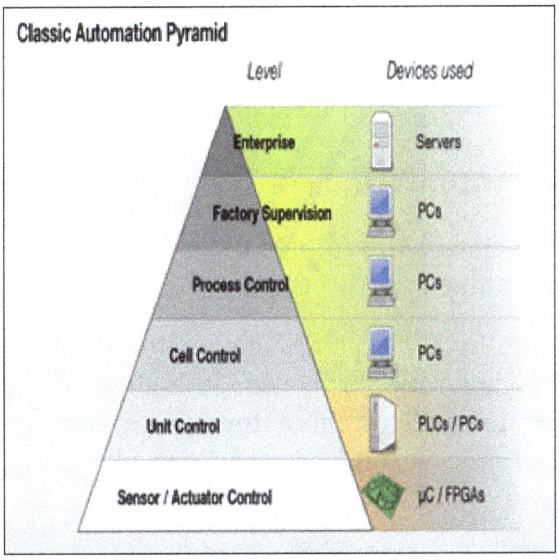

Automation Structure.

Sensor level is also called as process layer. It uses the sensors and actuators to get the values of the process variables in continuous or periodical manner. These act as eyes and arms of the industrial processes. Some of these instruments include pneumatic instruments, smart instruments, etc.

Automation control level or control layer uses industrial control devices like PC's/PLC's/DCS, etc. This level utilizes the various embedded processors, PID algorithms to control the process.

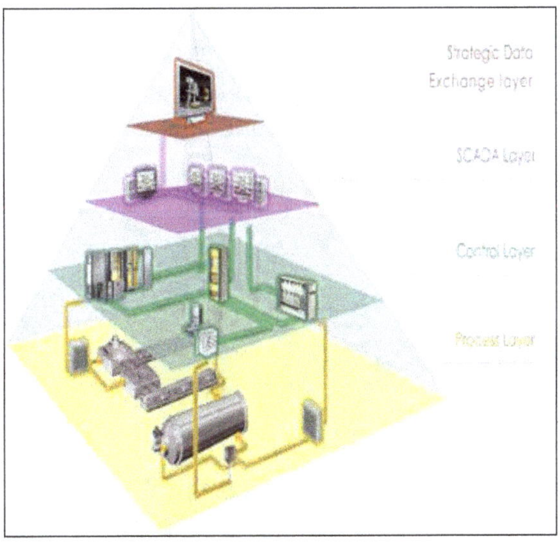

Automation Layers.

Supervising level or SCADA layer gets lots of channel information and stores the data in the system database. It acquires data from various control devices and displays them on HMI's (Human Machine Interface). It also gives alarm to indicate the levels of the process and control variables. It uses special software to get the data and communication protocols to interact with the field devices.

Enterprise level performs the tasks like scheduling, orders and sales, product planning, etc.

Types of Industrial Automations

Automation systems are classified into four basic types, based on the flexibility and level of integration in manufacturing processes.

- Fixed Automation: In this sequence of operations to be performed are fixed by the equipment configuration. It is used in high volume production with dedicated equipment. Examples of this automation system are automated assembly lines, distilled process, machine transfer lines.

Fixed Automation.

- Programmable Automation: In this, sequence of operations can be changed by changing the program. Sequence of operations is varied based on the different product configurations. Also new programs can be entered into the programmable devices for the new products. This type of system is used in batch processes, steel rolling mills, industrial robots, etc.

Programmable Automation.

- Flexible Automation: It is the extension for the programmable automation. This offers a greater flexibility to deal with product design variations. Operators can give commands in the form of codes in the computer program if wants to change the sequence of the process. Lower level equipments receive the instructions to operate at the field level without losing the production time. This type of automation is used in manufacturing multipurpose CNC machines, automatic guided vehicles, etc.

Flexible Automation.

- Integrated Automation: In this type total system is fully automated under computer control. Starting from designing process to the dispatching, whole system is completely automated. Even the equipment is handled by the robots. This system is used in computer integrated manufacturing systems.

Automation Equipments

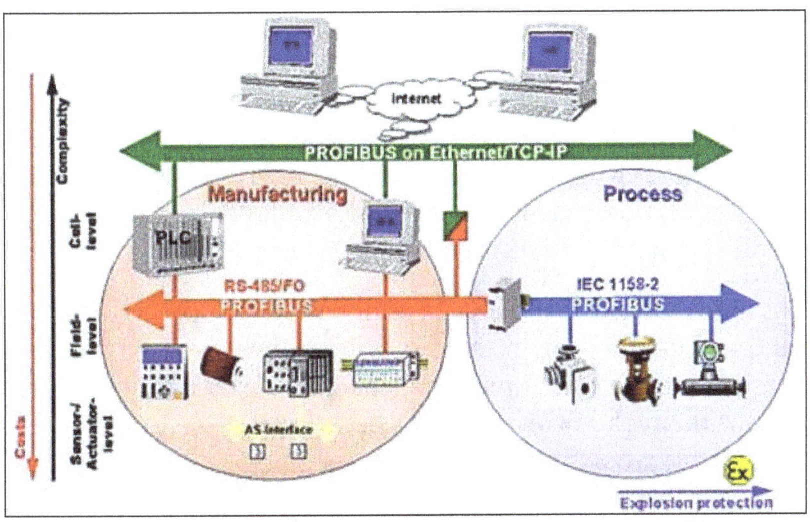

Equipments of Automated Industry.

- Sensors and Actuators: A sensor senses the various process variables and converts them into the electrical or optical signals. These sensors include temperature, pressure, velocity, flow, etc.

Actuators convert the electrical signals to the mechanical means to gain control over processes. These include the relays, magnets, servomotors, etc. Some of the sensors and actuators have the capability to communicate with the industrial field communication buses which comes under the smart devices.

- Industrial computers: Programmable Logic controllers (PLC's) also called as industrial computers are capable of being programmed to perform certain control functions. It consists of a CPU or processor, I/O modules (both analog and digital) to connect the various input/output devices and relay modules. These may be modular which is of fixed type or integrated types to extend modules based on the inputs available.

  Along with the PLC's, conventional PC's are used to control the process by online or by changing the programs. PLC's comes with dedicated software to program the control strategy.

- HMI(Human Machine Interface): HMI's offers the facilities like, displaying the information on computer screens and other displays, logging the results in the database, giving alarm signal, etc. It uses technologies like SCADA (Supervisory Control and Data Acquisition) and other visual based technologies.

- Communication system: In industries many sensors, actuators, controlling PC's and other control devices are geographically distributed and interacting with each other via several data buses. There are three types of buses used in the industrial automation i.e., factory bus, process bus, and the field bus.

Benefits of Industrial Automation

There are several reasons why companies opt for industrial automation systems:

- High productivity: Manufacturers can hire hundreds of workers and have them work around the clock to increase production levels. Even after doing that the plant still has to be closed for maintenance and repairs, and on holidays. So, it's not possible to have a 100% production up time.

  Industrial IoT automation helps to attain close to 100% up time by working 24 by 7 and 356 days a year. It leads to a dramatic increase in production levels. IIoT based predictive maintenance capability further helps to prepare companies deal with breakdowns in a much better way and reduce operational downtime.

- High Quality: Automated devices and robots don't tire or experience fatigues like humans do. They don't get bored and loose work focus. So, they don't make many errors.

  Industrial automation reduces errors and helps to enhance quality standards.

- High flexibility: Humans need training for new processes. Adding a new task in the assembly line results into many training hours for operators. Besides, operators have to get used to new processes which takes time. Companies have to suffer the loss incurred due to reduced production and trainer fees.

  Manufacturing robots and devices can be programmed to do varied tasks. New processes and methods can be instantly integrated in the manufacturing process. New production capabilities can be availed at affordable costs. This increases production flexibility.

- High information accuracy: Management needs production related data to make informed decisions. Data collection, if done manually, can prove to be costly. Moreover, there could be instances of human errors while collecting data. So manual data has to be verified and checked for reliability before it can be used for any meaningful purposes.

  The factory automation process gathers data automatically from sensors and devices. So it's not required to "sanitize" it. Moreover, data can be obtained frequently as and when required. Managements can acquire the most recent data before undertaking important decisions.

- High safety: Certain industrial environments are risky to work in. Many industrial processes use strong chemicals and heavy electrical voltage to manufacture goods. Workers and operators safety can be a concern in such cases. Companies have to provide, in some cases, large financial compensation in the event of work related accidents. Industrial automation systems make the production line safe for operators to work in. Robots handle all hazardous work.

Factory Automation Infrastructure

Factory automation infrastructure is a more recently coined phrase that is used to group the overall top-to-bottom end-to-end; products, procedures, building (or group of buildings), networks, controllers, interfaces, machinery, and assembly lines that are involved in the physical and organizational structures and facilities used in the production of goods. This grouping describes a general engineering and manufacturing environment that is defined by its ability to manufacture and assemble goods chiefly by machines, integrated assembly lines, and robotic arms. These environments are also defined by their coordination with (and usually their systematic integration with) the required automatic equipment to form a complete system. This system is essentially a solution for the automation and manufacturing of a particular production process of an intended output and end product.

Automation

Factory Automation with industrial robots for palletizing food products like bread and toast at a bakery.

The Automation process is a widely accepted form of production that has made great leaps and bounds in the production process as a whole. Automation has led to the production of sophisticated parts that have similar output qualities with little fluctuation in general tolerances. The use of automation in the manufacturing process started by using such technologies as the pneumatic and hydraulic systems for use in applications where their mechanical advantages could be used to utilize a production for higher output quality and efficiency. We have since created a complex and highly integrated systems composed of many different technologies and innovative procedures controlled under High Language programming environments with sophisticated operation drivers. These drivers are often running languages that support 6, 7, and 8 axis controls for sophisticated robotics.

Robotic Arm

6 Axis Articulated Robots from KUKA.

A robotic arm is a type of mechanical arm, usually programmable, with similar functions to a human arm; the arm may be the sum total of the mechanism or may be part of a more complex robot. The links of such a manipulator are connected by joints allowing either rotational motion (such as in an articulated robot) or transnational (linear) displacement. The links of the manipulator can be considered to form a kinematic chain. The terminus of the kinematic chain of the manipulator is called the end effect or and it is analogous to the human hand.

Advantages and Disadvantages

The main advantages of automation are:

- Increased through output or productivity.
- Improved quality or increased predictability of quality.
- Improved robustness (consistency), of processes or product.
- Increased consistency of output.
- Reduced direct human labor costs and expenses.
- The following methods are often employed to improve productivity, quality, or robustness.
- Install automation in operations to reduce cycle time.
- Install automation where a high degree of accuracy is required.
- Replacing human operators in tasks that involve hard physical or monotonous work.
- Replacing humans in tasks done in dangerous environments (i.e. fire, space, volcanoes, nuclear facilities, underwater, etc.).
- Performing tasks that are beyond human capabilities of size, weight, speed, endurance, etc.
- Economic improvement: Automation may improve in economy of enterprises, society or most of humanity. For example, when an enterprise invests in automation, technology recovers its investment; or when a state or country increases its income due to automation like Germany or Japan in the 20th Century.
- Reduces operation time and work handling time significantly.
- Frees up workers to take on other roles.
- Provides higher level jobs in the development, deployment, maintenance and running of the automated processes.

The main disadvantages of automation are:

- Security Threats/Vulnerability: An automated system may have a limited level of intelligence, and is therefore more susceptible to committing errors outside of its immediate scope of knowledge (e.g., it is typically unable to apply the rules of simple logic to general propositions).

- Unpredictable/excessive development costs: The research and development cost of automating a process may exceed the cost saved by the automation itself.
- High initial cost: The automation of a new product or plant typically requires a very large initial investment in comparison with the unit cost of the product, although the cost of automation may be spread among many products and over time.
- In manufacturing, the purpose of automation has shifted to issues broader than productivity, cost, and time.

Automatic Test Equipment

Automatic test equipment (ATE) is a machine that is designed to perform tests on different devices referred to as a devices under test (DUT). An ATE uses control systems and automated information technology to rapidly perform tests that measure and evaluate a DUT. ATE tests can be both simple and complex depending on the equipment tested. ATE testing is used in wireless communication and radar as well as electronic component manufacturing. There is also specialized semiconductor ATE for testing semiconductor devices.

Automatic test equipment is also known as automated test equipment. Automated test equipment is a computer-operated machine used to test devices for performance and capabilities. A device that is being tested is known as device under test (DUT). ATE can include testing for electronics, hardware, software, semiconductors or avionics. There are uncomplicated ATEs such as volt-ohm meters that measure resistance and voltage in PCs. There are also complex ATE systems that have several test mechanisms that automatically run high-level electronic diagnostics such as wafer testing for semiconductor device fabrication or for integrated circuits. Most high-tech ATE systems use automation to perform the test quickly.

The objective of ATE is to quickly confirm whether a DUT works and to find defects. This testing method saves on manufacturing costs and helps prevent a faulty device from entering the market. Because ATE is used in a wide array of DUTs, each testing has a different procedure. One actuality in all testing is that when the first out-of-tolerance value is detected, the testing stops and the DUT fails the evaluation.

Types of ATE Automatic Test Systems

There is a good variety of types of ATE systems that can be used. As they approach electronics test in slightly different ways they are normally suited to different stages in the production test cycle. The most widely used forms of ATE, automatic test equipment used today are listed below:

- PCB inspection systems: PCB inspection is a key element in any production process and particularly important where pick and place machines are involved. Manual inspection was used many years ago, but was always unreliable and inconsistent. Now with printed circuit boards that are considerably more complicated manual inspection is not a viable option. Accordingly automated systems are used:

- AOI, Automatic Optical Inspection is widely used in many manufacturing environments. It is essentially a form of inspection, but achieved automatically. This provides a much greater degree of repeatability and speed when compared to manual inspection. AOI, automatic optical inspection it is particularly useful when situated at the end of a line producing soldered boards. Here it can quickly locate production problems including solder defects as well as whether the correct components and fitted and also whether their orientation is correct. As AOI systems are generally located immediately after the PCB solder process, any solder process problems can be resolved quickly and before too many printed circuit boards are affected.
- AOI automatic optical inspection takes time to set up and for the test equipment to learn the board. Once set it can process boards very quickly and easily. It is ideal for high volume production. Although the level of manual intervention is low, it takes time to set up correctly, and there is a significant investment in the test system itself.
- Automated X-Ray inspection, AXI: Automated X-Ray inspection has many similarities to AOI. However with the advent of BGA packages it was necessary to be able to use a form of inspection that could view items not visible optically. Automated X-Ray inspection, AXI systems can look through IC packages and examine the solder joints underneath the package to evaluate the solder joints.

- ICT in circuit test: In-Circuit Test, ICT is a form of ATE that has been in use for many years and is a particularly effective form of printed circuit board test. This test technique not only looks at short circuits, open circuits, component values, but it also checks the operation of ICs.

  Although in Circuit Test, ICT is a very powerful tool, it is limited these days by lack of access to boards as a result of the high density of tracks and components in most designs. Pins for contact with the nodes have to be very accurately placed in view of the very fine pitches and may not always make good contact. In view of this and the increasing number of nodes being found on many boards today it is being used less than in previous years, although it is still widely used.

  A Manufacturing Defect Analyzer, MDA is another form of printed circuit board test and it is effectively a simplified form of ICT. However this form of printed circuit board test only tests for manufacturing defects looking at short circuits, open circuits and looks at some component values. As a result, the cost of these test systems is much lower than that of a full ICT, but the fault coverage is less.

- JTAG Boundary scan testing: Boundary scan is a form of testing that has come to the fore in recent years. Also known as JTAG, Joint Test Action Group, or by its standard IEEE 1149.1, boundary scan offers significant advantages over more traditional forms of testing and as such has become one of the major tools in automatic testing.

  The main reason that boundary scan testing was developed was to overcome the problems of lack of access to boards and integrated circuits for testing. Boundary scan overcomes this by having specific boundary scan registers in large integrated circuits. With the board set to a boundary scan mode, serial data registers in the integrated circuits have data passed into them. The response and hence data passing out of the serial data chain enables the tester to detect any failures. As a result of its ability to test boards and even ICs with very limited physical test access, Boundary Scan /JTAG has become very widely used.

- Functional testing: Functional test can be considered as any form of electronics testing that exercises the function of a circuit. There are a number of different approaches that can be adopted dependent upon the type of circuit (RF, digital, analogue, etc.), the degree of testing required. The main approaches are outlined below:

  - Functional Automatic Test Equipment, FATE: This term usually refers to the large functional automatic test equipment in a specially designed console. These automatic test equipment systems are generally used for testing digital boards but these days these large testers are not widely used. The increasing speeds at which many boards run these days cannot be accommodated on these testers where leads between the board under test and the tester measurement or stimulus point can result in large capacitances that slow the rate of operation down. In addition to this fixtures are expensive as is the programme development. Despite these drawbacks these testers may still be used in areas where production volumes are high and speeds not particularly high. They are generally used for testing digital boards.

  - Rack and stack test equipment using GPIB: One way in which boards, or units themselves can be tested is using a stack of remotely controlled test equipment.

    Despite its age, many items of rack mounted or bench test equipment still have a GPIB capability. Despite the fact that GPIB is relatively slow and has been in existence for over 30 years it is still widely used as it provides a very flexible method of test. The main drawback of GPIB is its speed and the cost of writing the programmes although test executive packages like LabView can be used to aid programme generation and execution in the test environment. Fixtures or test interfaces can also be expensive.

  - Chassis or rack based test equipment: One of the major drawbacks of the GPIB rack and stack automatic test equipment approach is that it occupies a large amount of space, and the operating speed is limited by the speed of the GPIB. To overcome these problems a variety of standards for systems contained within a chassis have been developed. Although there are a variety of ATE, automatic test equipment approaches that can be used, these are some of the more popular systems in use. They can all use test management software such as LabView to assist in the running of the individual tests. This enables facilities such as the ordering of tests, results collection and printout as well as results logging, etc.

- Combinational test: No single method of testing is able to provide a complete solution these days. To help overcome this various ATE automatic test equipment systems incorporate a variety of test approaches. These combinational testers are generally used for printed circuit board testing. By doing this, a single electronics test is able to gain a much greater level of access for the printed circuit board test, and the test coverage is much higher. Additionally a combinational tester is able to undertake a variety of different types of test without the need to mover the board from one tester to another. In this way a single suite of tests may include In-circuit testing as well as some functional tests and then some JTAG boundary scan testing.

Wireless Technologies in Factory Automation

Wireless technologies are utilised in consumer applications for several years, now. However, the operation of radio based communication systems in automation applications was considered doubtful for a long time. Primarily, the highly fluctuating quality of wireless transmission channels, compared to wired ones, was responsible for this fact. Transmitted electromagnetic waves experience reflexion, scattering, and diffraction, which may cause constructive or destructive interferences of the different signal copies arriving at the receiver. The direct consequences are packet errors and losses, resulting in higher transmission delays. Especially industrial propagation environments, with a lot of metallic surfaces and moving objects, are classified as demanding for wireless transmission. While fluctuations in latency and short losses of connections may be tolerated in consumer applications, exceeding given timelines in automation applications implies intolerable errors. The results are low plant availabilities and decreasing productivity.

By means of the development of diverse wireless standards and the adaption of well-suited protocols for industrial applications, the end-users doubts could be reduced dramatically over the last decay. The advantages of wireless solutions are obvious. In harsh environments, mobile and rotating scenarios, or at positions, difficult to access, cable connections and sliding contacts represent a main source of error. In this context the error probability can be decreased and the maintenance intervals increased by the utilisation of wireless technologies.

In addition to that, there is a great potential on saving time and money during planning, installation, and commissioning of plant sections. Many domains of the industrial automation already profit from the deployment of wireless solutions. With KNX RF and ZigBee the first standards for building automation have been introduced. Within the scope of the HART 7 specifications the first standard, WirelessHART, for the process automation was released in late 2007. Further standards, especially for the domain of factory automation, are expected to get published in 2010. In order to reduce costs and time during the development of wireless solutions, unlicensed frequency bands are typically used for operation. Moreover an almost worldwide harmonised operation is guaranteed. This tendency is very pronounced for the 2.45GHz ISM (Industrial, Scientific, and Medical) frequency band. Because of the high availability, transceiver chips of commercial standards are often applied on the physical layer. Good examples are the technologies of IEEE 802.11, IEEE 802.15.1 Bluetooth or IEEE 802.15.4. Previous analysis attested the capabilities of these technologies for industrial applications. In order to comply with the requirements in automation with respect to determinism, reliability, and availability, above the physical layer adapted protocols are implemented. With latency requirements of 100ms <1ms, the domain of factory automation is one of the most demanding. With state of the art technologies, wireless solutions may serve applications with update times of 10ms...20ms. Below these timelines the reserve for packet retransmissions and error corrections of the current technologies is insufficient to guarantee a reliable communication. The upcoming ultra wideband (UWB) technologies are considered potential candidates for faster wireless solutions of the next generation.

Performance Requirements of Communication Systems in Factory Automation

The hierarchical model of automation. With respect to the current state of the art, wireless technologies may be applied, as a supplement to existing wired communication networks, across all levels of the model. The major part of automation devices is equipped on the control and sensor/actuator level.

Basically, the control level consists of programmable logic controllers (PLCs) and (intelligent) decentralised I/O-Terminals, which are interconnected by one or more field buses. Depending on the plant, the number of devices varies and their distances may range from a few to some hundreds of meters. Classical field buses for example are Profibus or Interbus. Current field buses aim at convergence to the well-established Ethernet IT-networks with data rates of up to 100Mbps. Popular examples are Profinet, EtherCat, Ethernet-Powerlink, Sercos III and Ethernet/IP. Depending on the area of application typical cycle times of 10ms...100ms (PLC to PLC, Visualization), 1ms...10ms (Communication to decentralized peripherals), and <1ms (Motion Control) are required.

The transition from control level to the sensor/actuator level is fluent. At this level sensors, actuators, drives, and manufacturing robots are equipped. The number of I/O-points is very high and densities of $2/m^3$ to $10/m^3$ can be expected in typical production plants. The major part consists of sensors like proximity switches. The connection of sensors and actuators to PLCs can be realised in different ways. A differentiation is made between parallel and serial wiring. The classical parallel wiring uses analogue or digital (remote) I/O-Terminals to connect the sensors and actuators with the field bus system. New concepts, like IO-Link, use a 2/3-wire-standard-parallel-wiring of the sensors and actuators, as well. However, on the basis of a master/slave protocol, additional information, like diagnosis or parameters for configuration, can be exchanged with a PLC besides the process data. IO-Link specifies data rates of at least 4.8kBaud and 38.4kBaud. Corresponding to the amount of process data, cycle times of 2ms can be achieved at data rates of 38.4kBaud. The serial wiring is usually realised by means of a sub bus system, which serially interconnects the sensors and actuators. The most representative technology is the AS-Interface. At a maximal amount of 31 (62) devices, the cycle times are <5ms (10ms). Because of the high requirements concerning latency, drives and robots are interfaced directly to the field bus via appropriate application profiles or special drive buses, i.e. Sercos.

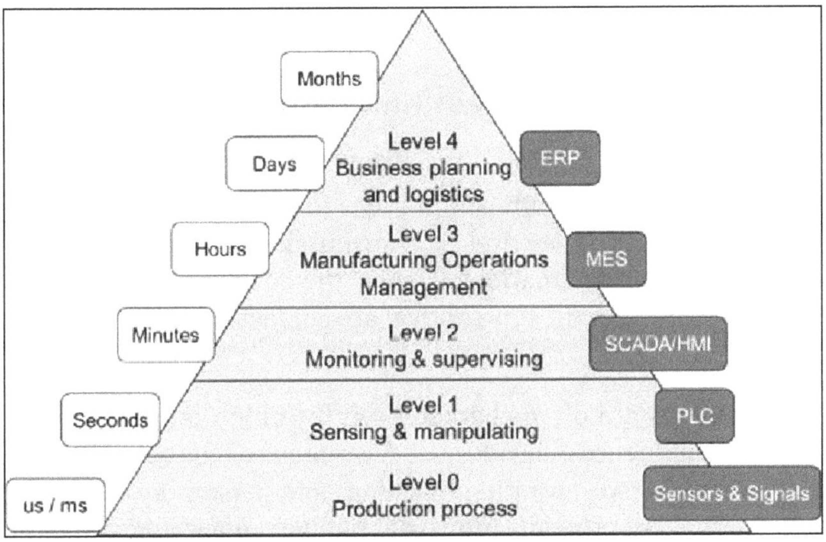

The hierarchical automation pyramid.

The application of wireless technologies is useful on the control as well as the sensor/actuator level. However, besides the stringent requirements concerning latency, determinism, reliability, and data integrity, the requirements with respect to costs, energy consumption, distances, bandwidth, node density, and network topologies differ.

Depending on the application, the requirements with respect to costs, energy consumption and node density are of minor importance on the control level. Because of the wide spreading of PLCs and decentralised peripherals, the ranges between devices may vary from a few meters up to some hundreds of meters. The process data of a whole plant is exchanged via the field bus system. In this conjunction the amount of process data of decentralised peripherals may vary between <10 Bytes...>100Bytes. Hence, the bandwidth and data rates are of major importance. The size of the actual data packets depends on the structure of the field bus system and whether it uses multi-slave or single-slave frames. The network topologies for wireless solutions range from simple cable replacement point-to-point and point-to-multipoint connections up to cellular networks with roaming capabilities (production lines, automated guided vehicles).

Because of the high quantities of devices, the costs for acquisition, installation, commissioning, and operation are of major importance on the sensor/actuator level. The sphere of action is restricted to small production cells (10m^3...100m^3) with high node densities. The amount of process data of a single sensor or actuator typically ranges from 1Byte...10Bytes. Hence, lower data rates and bandwidth are sufficient. In its simplest form, wireless solutions operate as cable replacements in point-to-point topology, as well. However, the development is focused on high speed wireless sensor/actuator networks (WSANs), supporting large numbers of devices. These networks are usually arranged in star topology and consist of wireless sensors/actuators, wireless I/O-concentrators, and a master base station, which acts as the interface to a super ordinate control system. Due to the increasing latencies, multihop topologies are currently not considered for WSANs in factory automation.

Industrial Wireless Communication Channels

Communication systems have to comply with the stringent requirements concerning reliability, availability, and determinism in order to serve automation applications. In contrast to that, the quality of a wireless transmission channel experiences random time and frequency variant fluctuations. Hence, the development of wireless communication systems, for the extreme time critical area of factory automation, is a big challenge.

Industrial environments are often characterised by a high degree of metallic surfaces and time-varying influences. Besides the movement of the radio systems itself the movements of materials/tools, rotating machines and persons are responsible for this time variant properties. In principle industrial radio channels are akin to mobile radio channels. Thus, most phenomena of industrial radio channels comply with the ones of mobile radio channels. The occurring physical phenomena of transmitted electromagnetic (EM) waves are illustrated in figure:

- Reflexions occur, when EM-waves encounter reflecting objects, whose dimensions are much larger than the wavelength.
- Scattering appears, either when the dimensions of the encountered object are much smaller than the wavelength of the EM-wave, or when the surface structure is classified very rough in comparison to the wavelength.
- Diffraction occurs when EM-waves encounter sharp edges.
- Shadowing is caused by obstacles, which completely block the propagation paths of EM-waves.

- Doppler effects arise, either when there is a relative movement between transmitter and receiver, or a mobile obstacle in the propagation field reflects, scatters, diffracts, or shadows the EM-wave.

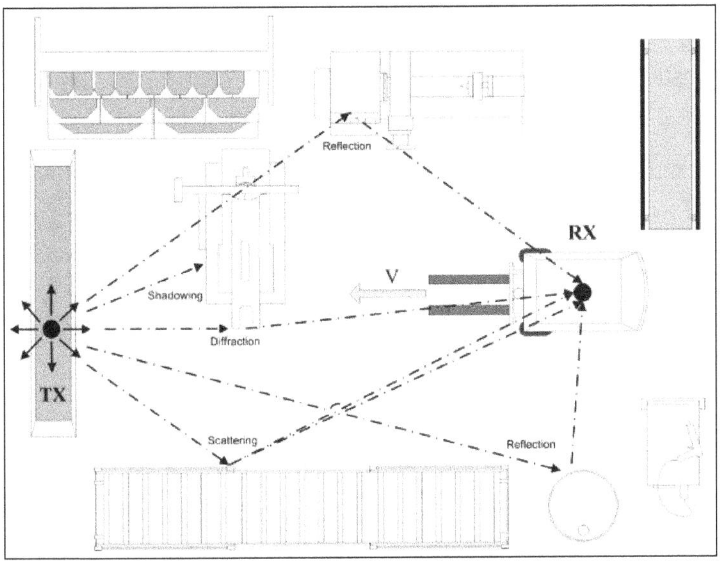

The classical propagation of electromagnetic waves in a typical industrial environment.

Because of these wave phenomena a received signal is a composition of different attenuated and phase shifted versions of the original transmitted signal. Depending on the phase of these versions, a constructive or destructive overlapping occurs at the receiver. This effect is called multipath scattering. The absence of a direct non reflected version of the transmitted signal is typical for industrial radio channels. Does a relative proper motion between the transmitter and receiver take additionally place, or does the environment change due to rotating machines or forklift trucks, a shift in frequency based on the doppler effect influences the transmitted signal. Simultaneously, the path of the signal versions change, resulting in a new form of the received signal. Hence, the transmission behaviour of such a radio channel is time-variant and the signal power experiences high fluctuations.

Large Scale Fading

The large scale fading results from widespread movements. It depicts the mean signal power over spatial areas of about 10 wavelengths λ. Consequently, the local mean values of the propagation losses (path loss), which depend on the environment (shadowing, reflexion, diffraction, scattering), are characterised. In this conjunction the log-distance path loss model is often used to describe path losses. The model states, that the mean received power *Pr* decreases logarithmical with the distance *d* between transmitter and receiver, following $Pr \propto P_0 d_{0d} \gamma$. d_0 is a reference distance near the transmitter, where the transmit power P_0 is measured with respect to the far-field characteristics of the transmit antenna. The degree of signal attenuation is expressed by the path loss exponent γ. In buildings γ may vary very much. At frequencies of 400MHz...4GHz γ can take values of γ=2,...,6. In analysis of Rappaport, performed in five different factory environments, mean values of γ=1.7,...,3 were measured.

Small Scale Fading

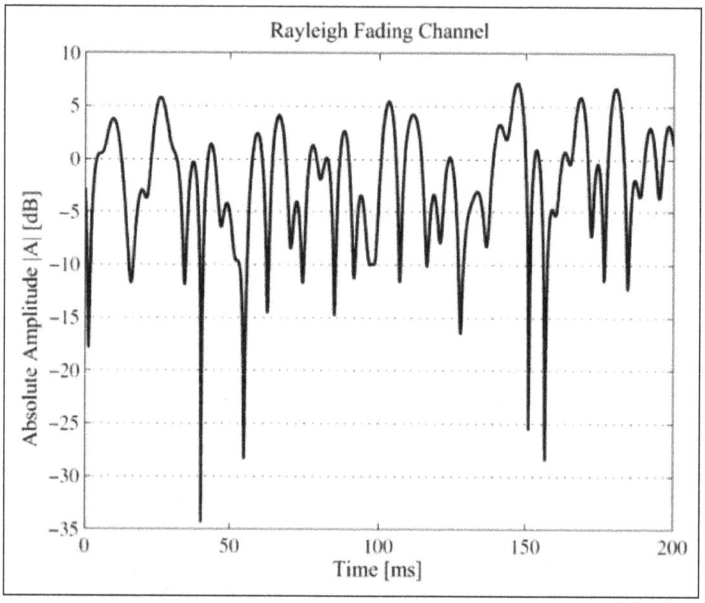

The course of amplitudes of a rayleigh fading channel.

Small scale fading characterises the fast fluctuations of radio channels over short distances (fraction λ). Primarily, these fast fluctuations of the channel are caused by doppler effects and multipath scattering. If, for example, a narrow band carrier signal is transmitted, several randomly organised signal copies arrive at the receiving antenna via different paths. For every location in a propagation environment, the received signal is the sum of all signal versions. If the signal versions, which arrive at the receiver, are uncorrelated in phase, the angles of arrival uniformly distributed, and the signal delay of each path much lower than the alteration speed of the radio channel, then the behaviour of attenuation can be described by two complex gaussian processes with mean values of $\mu = \mu R + j \cdot \mu I$. If there is no direct line of sight (NLOS) between transmitter and receiver, the mean value is μ=0. In this case the probability distribution of the absolute amplitude values corresponds to the rayleigh distribution. If there is a direct line of sight (LOS), the mean value μ takes the amplitude value of the signal version, transmitted over the direct path $\mu=ALOS$. The absolute amplitude values of these channels correspond to the rice distribution. Figure shows a classical course of the absolute amplitude values of a rayleigh fading channel. The deep fades of up to 40dB are characteristic. Analysis in industrial environments showed a dynamic range of 20dB in signal power, for stationary transmitters and receivers. When the receiver was moved with a velocity of v=0.3ms, the dynamic range of the received signal increased to 30dB...40dB. If a channel experiences such a deep fade, several channel errors occur, whose positions show a strong statistical dependence. The occurrence of channel errors temporarily appears in complex blocks.

Since the rayleigh and the rice models are derived on the assumption of a non-modulated carrier signal, their application is restricted to narrow band signals.

In order to completely characterise a radio channel with respect to the domains of time and frequency, the time variant impulse response h(τ,t) is an appropriate measure. On the supposition of a wide sense stationary uncorrelated scattering (WSSUS) channel, the following characteristics can be approximated on the basis of Fourier transformations of h(τ,t) and the computation of first and second order statistics:

- Delay Spread: The delay spread τrms describes the mean spread in time of transmitted δ-impulse. Scientific studies showed a delay spread of τrms=20,...,30ns at frequencies of 1.3GHz in industrial environments. In this conjunction the works of Haehniche et al. are of great practical interest. The delay spread for the 2.45GHz ISM frequency band was analysed in different industrial environments. A mean value of 72ns and a maximal value of 121ns were measured. Hoeing et al. analysed the delay spread in a production cell with several scattering obstacles. The transmission distance was 3m with LOS between transmitter and receiver. Within the propagation area of interest, fast cyclic movements of machines took place. Under these conditions a delay spread of τrms=79ns was measured, which corresponds to a path difference of about 23.7m in length.

- Coherence Bandwidth: Within a frequency area of Δf, which is smaller than the coherence bandwidth BC, the course of amplitudes is expected to be constant. Between the delay spread and the coherence bandwidth the approximation $\tau rms \approx BC-1$ is valid. Haehniche et al. analysed the coherence bandwidth in different industrial environments, as well. Mean values of the coherence bandwidth BC=5.7MHz were measured for the 2.45GHz frequency band. In a coherence bandwidth of up to 10MHz is reported for this frequency range.

- Coherence Time: The coherence time TC is a measure for a radio channels alteration speed.

- Doppler Spread: The doppler spread DS describes the mean frequency spread of a tranmitted narrow band carrier signal. Between the coherence time and the doppler spread the approximation $TC \approx DS-1$ is valid. The impact of the doppler spread in industrial radio channels may be enorm. Fast moving or rotating machines may induce high values of the doppler spread. Hoeing et al. have measured values for DS of up to 400Hz.

On the basis of the presented characteristics, the small scale fading can be further classified with respect to the variance in time and frequency of a radio channel. If the signal bandwidth is much smaller than the coherence bandwidth $BS \ll BC$, and the delay spread much smaller than the symbol duration $\tau rms \ll TS$, the radio channel is characterised as flat fading (non-frequency selective). Flat fading channels are often referred to as narrow band channels. If the signal bandwidth is larger than the coherence bandwidth $BS \gg BC$, the channel is frequency selective. In this case the delay τ of single paths is larger than the symbol duration TS, what might induce intersymbol interferences (ISI) at the receiver. The time selectivity of a radio channel may either be described on the basis of the coherence time TC or the doppler spread DS. If the symbol duration is much smaller than the coherence time $TS \ll TC$, the form of the transmitted symbol is not altered by the radio channel. These channels are referred as slow fading (non-time selective). The opposite is a time selective radio channel referred to as fast fading.

Performance-enhancing Strategies

In order to comply with the challenging requirements of automation in the face of the depicted fluctuations of industrial radio channels, several performance enhancing strategies can be applied. It is obvious, that these methods are most effective, when implemented in the PHY or MAC layers. However, with the given architectures of available transceivers it is often necessary and only possible to implement appropriate protocols on application layer.

Classical methods to improve the performance of radio channels are error detecting (retransmissions) or error correcting codes, which add further redundancy to the transmitted data. Since these methods are typically applied to a single channel, their effectiveness mostly depends on the small scale properties of the channel. Deep fades induce dense blocks of errors, which can be hardly corrected by error correcting codes. The success of a retransmitted signal depends on the duration of these deep fading (coherence time). A way to overcome these problems is the utilisation of diversity techniques. In general diversity describes the transmission of information over different channels. The achievable gain depends on the statistical independence of each transmission channel. With an increasing number of independent transmission channels the probability increases, that at least one channel is in a good state, and the transmitted signal can be decoded at the receiver. If the error generating processes are completely uncorrelated, the theoretical minimal error probability is $Pr=Pen$ for n transmission channels. Diversity techniques can be applied in the domains of time, frequency, space and angle. Since time diversity implies an increasing latency, its operation in time critical applications is not suitable. However, by applying spatial or frequency diversity, significant gains at reasonable costs can be achieved.

Spatial diversity may be applied in different forms. A classification is made for single-user and multi-user approaches. In the case of single-user, there is only one transmitter and one receiver, with at least one of which having multiple antennas.

In it is proven, that the achievable capacity nearly linearly increases with $N \rightarrow \infty$, if both transmitter and receiver are equipped with the same number of antennas N. In its simplest form, multiple antennas are used at the receiver (SIMO). The single signal versions are combined at the receiver in order to produce the received signal. Well known combining techniques are switched combining, equal gain combining or maximum ratio combining. The achievable diversity gain thereby depends on the statistical independence of the received signals. On the assumption of a rayleigh fading channel the normalised correlation coefficients $\rho\zeta$ of two envelopes can be expressed as a function of antenna separation $\rho\zeta = J_{02} \cdot (2\pi\zeta)$. ζ represents the seperation of two vertical monopole antennas in wavelengths and J_0 is the Bessel function of first kind and zero order. In practical measurements, in order to evaluate digital diversity techniques, were performed, based on a multi-transceiver platform, operating in the 2.45GHz frequency band. By utilising three receiving antennas at a separation of 4.69cm a diversity gain of 3.5 dB could be realised in an industrial environment. Bit error rates (BER) could be reduced by half an order of magnitude compared to a single branch. The packet error rate (PER) could even be reduced by more than one order of magnitude. Based on more complex MIMO approaches, i.e. applied in the upcoming standard IEEE 802.11n, performance gains can be further increased. The capabilities of multi-user approaches, i.e. relaying, for industrial applications has been demonstrated in.

A second form of diversity is the transmission of Information over multiple frequencies. The achievable diversity gains depend on the statistical independence of the single transmission channels, as well. To obtain statistical independence between two channels their frequency separation should at least be larger than the actual coherence bandwidth. Following, the normalised correlation coefficient $\rho\Delta f$ of two envolpes can be expressed as a function of frequency separation $\rho\Delta f = 1 + (2\pi T\Delta f) - 12$. Thereby Δf describes the separation of the two frequencies and T is the maximal delay spread of a current environment. In narrow band systems frequency diversity is often combined with time diversity in the form of "frequency hopping spread spectrum" (FHSS). In wide band systems, which use "orthogonal frequency division multiplex" (OFDM), frequency diversity is often applied on the basis of channel coding combined with interleaving in the frequency domain. In the performance of frequency diversity at frequencies of 1.75GHz...1.8GHz has been evaluated in typical office buildings. At an availability of 99%, the achieved diversity gains varied between 5dB...9.6dB for frequency separations larger than 5MHz.

Having in mind the limitation of bandwidth and consumption of energy, spatial diversity is the more attractive strategy. However, frequency diversity is also considered a suitable instrument to compensate deep fading. Although it is proven, that optimum combining, using spatial diversity, may increase the signal to noise plus interference ration (SINR) in order to mitigate co-channel interferences, the application of frequency diversity is more effective and less complex.

Current Wireless Base Technologies and its Utilisation in Factory Automation

Most of the industrial wireless solutions use the unlicensed 2.45GHz ISM frequency band.

Regulation for the 2.4 GHz ISM Frequency Band

Within the scope of the regulation 5.138 and 5.150 of the international telecommunication union, radio communication sector (ITU-R), besides others, the frequency range from 2.4GHz to 2.5GHz is enabled for industrial, scientific, and medical (ISM) applications. The European norm EN 300 328 regulates the frequency range from 2.4GHz to 2.4835GHz for general utilisation in Europe. The maximal EIRP transmit power is limited to 100mW. For devices, that do not use the modulation of "frequency hopping spread spectrum" (FHSS), the maximal spectral EIRP power density is further limited to 10mW/MHz. There are no restrictions concerning the duty cycle of the radios. Depending on the application domain and the country, transmit powers above 10mW have to be registered. In general, there are country specific limitations to the utilisation of the 2.45GHz ISM band (i.e. Spain and France).

In North America, the utilisation of unlicensed frequency bands is ruled by the Federal Communications Commission in the document CFR 47, Part 15. The maximal transmit power for the 2.45GHz band is limited to 1W for systems using FHSS over more than 75 frequency channels. For systems with less than 75 channels, the maximal transmit power is limited to 125mW. In addition to that, a spectral power density of 8dBm/3kHz must not be exceeded.

Wireless Local Area Networks - IEEE 802.11

The most popular radio technologies operating within the 2.45GHz band are compliant to the standards of IEEE 802.11b and IEEE 802.11g. Both standards specify 13 channels with spacing of 5MHz for Europe and 11 for North America.

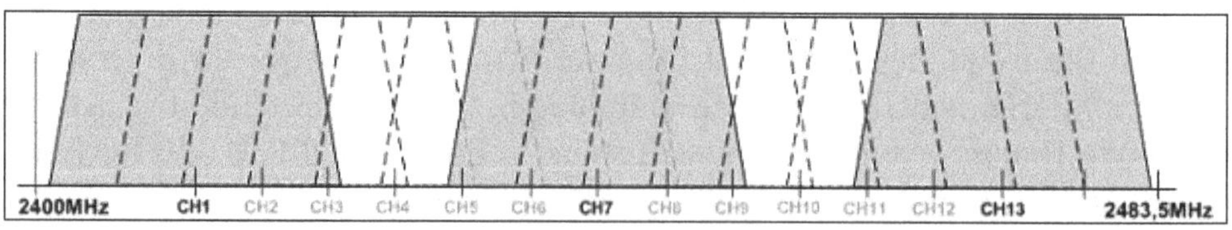

IEEE 802.11 defines 13 channels for Europe and 14 Channels for North America.

With a transmit bandwidth of about 20MHz, three non-overlapping channels with a spacing of 30MHz are available. The maximal transmit power is limited to 100mW.

IEEE 802.11b supports data rates of 1Mbps...11Mbps. According to the selected data rates, the modulations of "differential binary phase shift keying" (DBPSK), differential quadrature phase shift keying (DQPSK) or, complementary code keying (CCK) are used. "Direct sequence spread spectrum" (DSSS) is used as a spreading technique. The amendment of IEEE 802.11g is an extension and supports data rates of up to 54Mbps by introducing "orthogonal frequency division multiplex" (OFDM) with 52 sub-carriers as a spreading technique. These sub-carriers are either modulated using binary phase shift keying (BPSK), quadrature phase shift keying (QPSK), 16- or 64-quadrature amplitude modulation (16-QAM, 64-QAM) depending on the selected data rates. Furthermore this standard supports forward error correction (FEC) with coding rates of 1/2, 2/3, or 3/4. As the channel access method, both standards use "carrier sense multiple access/collision avoidance", which is based on a "clear channel assessment" (CCA) module. Prior to any transmission, the CCA module validates the occupation of the medium. If the medium is classified "busy", the transmit operation is interrupted for a pseudo random period of time and the channel is validated again. A prioritised medium access, comprising eight priority levels, was introduced by the extension of IEEE 802.11e. In order to classify the medium, three modes are specified and one of them must at least be supported. In mode 1 the medium is considered busy, as soon as the detected energy is above a predefined threshold. In mode 2 the medium is considered busy, if an IEEE 802.11 modulated signal is detected. In mode 3 the medium is considered busy, if an IEEE 802.11 modulated signal is detected and its energy is above a predefined threshold. In general, the end-user has no access to the configuration of the CCA mode.

In automation applications IEEE 802.11 is recommended by the PROFIBUS & PROFINET International (PI) as a wireless communication system for connecting PLCs and decentralised peripherals. With adapted IEEE 802.11 systems, PROFINET-I/O communications with update times of up to 8ms can be served. Common use cases are forklift trucks and automated guided vehicles. In mobile scenarios the transition from one cell to another (roaming) is extremely critical. Currently, roaming times of <50% can be realised.

The next Amendment of the task group IEEE 802.11n is shortly before being published. This standard specifies either channels with 20MHz bandwidth and 56 OFDM sub-carriers and channels with 40MHz bandwidth and 112 sub-carriers within the frequency bands of 2.45GHz and 5GHz. By applying performance enhancing techniques like "MIMO", "Channel Bonding", "Frame Aggregation", "Spatial Multiplexing", and "Beam forming", data rates of 300Mbps and beyond can be achieved. At the moment the draft standard, revision 8, is available. The release of the final standard is expected in late 2009. Similar to the standards IEEE 802.11b and IEEE 802.11g a fast market penetration can be expected for the standard IEEE 802.11n, as well.

Bluetooth – IEEE 802.15.1

The latest specification of Bluetooth version 3.0 was published in 2009. The PHY and MAC layer of the Bluetooth version 1.1 are published as the standard IEEE 802.15.1, as well. In its classical form 79 channels, with a spacing of 1MHz, are specified in the range of 2.402GHz...2.480GHz. The radio signals are modulated using "Gaussian frequency shift keying" (GFSK, 1 Mbps), "π/4 differential quaternary phase shift keying" (π/4-DQPSK, 2 Mbps), or "8-ary differential encoded phase shift keying" (8DPSK, 3 Mbps). Bluetooth uses "Time Division Multiple Access" (TDMA) as the channel access method and FHSS for spreading. Three device classes with transmit powers of 1mW, 2.5mW and 100mW are defined.

Bluetooth networks, called piconets, are formed in star topology. A piconet consists of a master and up to seven active slaves. In order to communicate, timeslots with a length of 625µs are predefined. The specification defines synchronous connections (SCO) for the transmission of i.e. speech and asynchronous connections (ACL) for data transmission. Depending on the type, data packets occupy one to five timeslots and use "automated repeat requests" (ARQ) or FEC as channel coding. In each timeslot, or at least after the transmission of a data packet, a change in frequency is performed respectively.

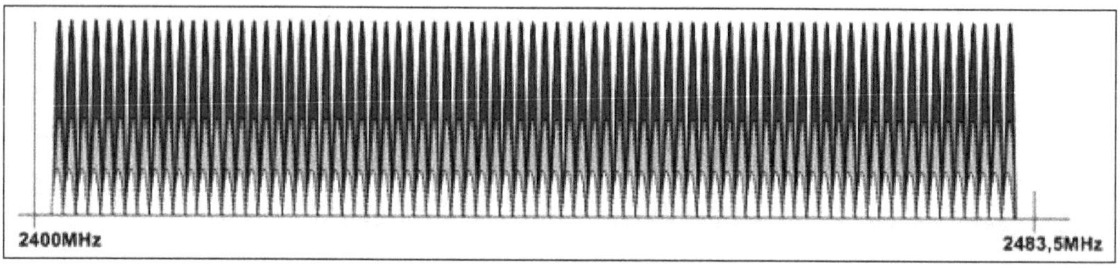

IEEE 802.15.1 defines 79 Channels within the 2.45GHz ISM Band.

In avoidance of coexistence problems, the standard supports an "adaptive power control" (APC) and "adaptive frequency hopping" (AFH). When using AFH, frequency channels occupied by foreign radios are detected and excluded from the hopping scheme. With common Bluetooth transceiver chips a channel is classified busy, when the occupation is higher than 15%. The adaption of the hopping scheme depends on the implementation and may take up to several seconds. In addition to the adaptive channel classification, frequency channels can be excluded of the hopping scheme manually, in order to avoid frequencies known to be in use by other radios. At least 20 channels have to be used. By doing so, a frequency separation to two coexisting IEEE 802.11 radios can be administered. Solely, the connection setup uses all frequencies. However, some vendors developed standard compliant solutions, which prevent interferences during the connection setup.

Bluetooth is applicable at control as well as sensor/actuator level. With respect to ABBs "Wireless interface for sensors and actuators" (WISA), the PROFIBUS & PROFINET International (PI) actually considers the PHY layer of Bluetooth as the basis for "Wireless Sensor/Actor Networks" (WSANs). A standard shall be published in 2010. A WISA network consists of a base station and up to 120 wireless I/O-concentrators and sensors/actuators in a star topology. The base station acts as the network coordinator and gateway to a super ordinate control system. The I/O-concentrators and sensors/actuators use IEEE 802.15.1 standard compliant transceivers. The base station consists of a special multi-transceiver architecture and thus able to serve multiple devices in parallel. The update time of 120 sensors is typically below 20ms.

In version 3.0 of Bluetooth, the support of IEEE 802.11 as an "Alternate MAC PHY" (AMP) is introduced. In addition to that the "Bluetooth Low Energy" specification is to be published in late 2009. First transceivers for both technologies shall be available in 2010.

IEEE 802.15.4

The standard IEEE 802.15.4 specifies 16 channels with a separation of 5MHz for the 2.45GHz ISM band. With DSSS as spreading and "offset quadrature phase shift keying" (O-QPSK) as modulation, data rates of 250kbps are supported. The standard limits the transmit power to 1mW. However, the regulations allow the operation at transmit powers of up to 10mW.

As channel access method CSMA/CA corresponding to IEEE 802.11 is utilised. Optionally, the standard supports a synchronised data communication in superframes of durations from 15ms to 246s. Each superframe consists of a "contention access period" (CAP) and a "contention free period" (CFP). During the CAP, devices willing to transmit, concurrently access the medium via CSMA/CA. The CFP consists of guaranteed timeslots and gives exclusive access to medium for higher prioritised transmissions. The standard was designed for low power industrial "wireless personal area networks" with low data rates.

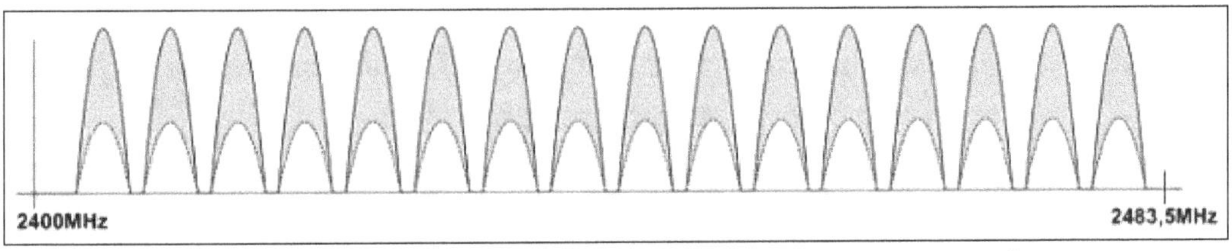

IEEE 802.15.4 defines 16 Channels within the 2.45GHz ISM Band.

The technology is wide spread in combination with the higher layers specified by ZigBee. ZigBee supports the operation of large multihop networks and addresses domains like home- and building automation, smart metering, and health care.

Within the scope of the HART 7 specifications, the first wireless standard for process automation, WirelessHART, was published in 2007. WirelessHART is based on the PHY layer of IEEE 802.15.4 and uses the "Time Synchronized Mesh Protocol" (TSMP) for channel access. In order to improve reliability, it is designed to support large multihop networks in full mesh topologies with a high degree of redundant paths. In avoidance of coexistence problems the standard changes frequencies at a rate of 10ms. Optionally, a channel black list can be used to avoid frequencies currently in use. First products are successfully in use since late 2008.

At the moment "the International Society of Automation" (ISA) is shortly before publishing a second standard for the process automation, ISA 100.11a, based on the PHY layer of IEEE 802.15.4. In the domain of factory automation a few proprietary solutions for the transmission of sensor data based on IEEE 802.15.4 are available. Right now the task group of IEEE 802.15.4e is working on MAC layer extensions. In order to improve the support of time critical industrial applications, shorter transmit times, improved TDMA techniques and frequency hopping are evaluated. In the long run the extensions of IEEE 802.15.4e shall enable the standard to better support applications in factory automation.

Coexistence in the 2.4 GHz ISM Frequency Band

With the fast pace growth of wireless solutions, operating in the 2.45GHz ISM band, in automation as well as the IT, the end-users demand for a good coexistence of the devices is getting obvious. In this respect a technologies coexistence properties depend on several parameters, like the transmit power, signal bandwidth, channel access methods, and duty-cycle, which often are vendor specific.

In IEEE 802.15.2 coexistence is defined as "a systems ability to perform a task in a shared medium, while other systems perform their tasks, complying with the same or a different set of rules". In a shared medium the main source of error is caused by interferences. Interferences appear, when signals overlay in the domains of time, frequency, and space. For the domain of frequency the IEEE Unapproved Draft Std P1900.2/D2.22 further subdivides interferences into "In-Band", consisting of "Co-Channel-" and "Adjacent Channel- Interference", and "Out of Band", consisting of "Band Edge-" und "Far out of Band Interference". The most common form of appearance are "Co-Channel" interferences, which occur, when more than one system operates on the same frequency.

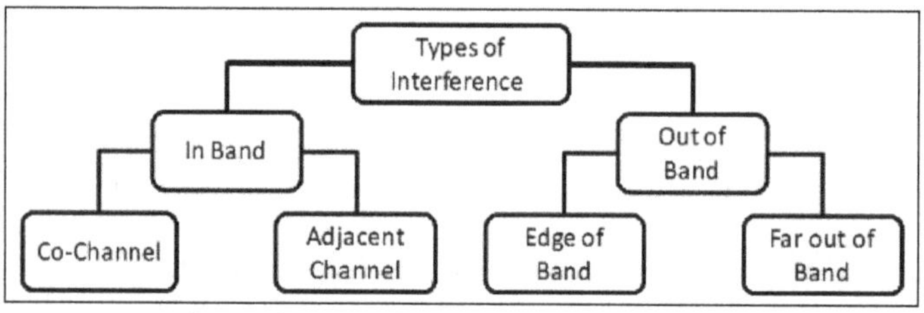

Types of Interference defined by IEEE P1900.2/D2.22.

The domain of time is determined by the channel occupation in time, the duty cycle, of coexisting systems. The probability of signal interferences increases with the utilisation of the medium in time. The spatial domain is defined by the transmit power, the distance between the systems (Antennas), and the resulting "signal to interference plus noise ratio" (SINR). If the SINR is too low, a signal cannot be detected correctly at the receiver.

Besides these physical properties of interferences, channel access methods have a strong impact on the coexistence of radios. Typically, radio systems operating in the 2.45GHz ISM band use either TDMA, CSMA/CA, or a mixture of both as access methods. TDMA subdivides the medium into timeslots, which are reserved for exclusive access to the medium. That way, TDMA systems support a deterministic behaviour in time and a good coexistence within the same network. In order to avoid interferences to foreign networks, TDMA is often used in combination with FHSS, additionally allowing to black list frequencies already in use by other systems (i.e. Bluetooth). When using CSMA/CA, the state of the medium is validated before any transmission of data and only performed, if the medium is classified idle. The validation of the medium is either based on an energy threshold, the detection of a valid carrier, or a mixture of both. On the one hand CSMA/CA is able to avoid interferences within the same or foreign networks. On the other hand CSMA/CA is vulnerable to jamming attacks and some kind of unnecessary interferences. Depending on the implementation, the following types of interferences may occur, when using CSMA/CA:

- Type-1: A weak signal, that would not induce interferences at the receiver, is detected at the transmitter, causes the medium to be classified busy, and thus delays the transmission ("Exposed Terminal Problem").
- Type-2: Interferences caused by multiple radios that access the medium at the same time.
- Type-3: The source of interference is out of the detection range of the transmitter, but causes interferences at the receiver ("Hidden Terminal Problem").

There are several strategies to mitigate these interferences within the same network of operation. However, interferences with foreign networks may still appear. How far interferences actually influence the coexistence properties of a system, always depends on the tasks to be performed. Usually, an underlying (wireless) communication system has a temporal reserve with respect to an application, in order to perform channel coding and retransmissions. If this reserve gets exhausted, the communication system cannot longer serve the application. It is obvious, that with increasing temporal requirements of an application, the reserve of the communication system decreases and interferences result in application errors faster. Analytical as well as practical studies about the coexistence within the 2.45GHz ISM band have been subject to several publications.

The previous descriptions stated the richness of technologies and applications operating in the 2.45GHz ISM band. Thus, a coexisting operation of different wireless solutions is hardly avoidable. But it is very demanding to consider all parameter of relevance for the different domains of applications, when determining the properties of coexistence of radio technologies. In addition to that, comprehensive studies on the coexistence of new technologies, like IEEE 802.11n, WirelessHART, ISA 100.11a, and Bluetooth Low Energy have not been performed, yet.

Table: Application classes of ISA-SP100.

Category	Class	Application	Description
Safety	0	Emergency action	(always critical)
Control	1	Closed loop regulatory control	(often critical)
	2	Closed loop supervisory control	(usually non-critical)
	3	Open loop control	(human in the loop)

Monitoring	4	Alerting	Short-term operational consequences (e.g. event-based maintenance)
	5	Logging and downloading/uploading	No immediate operational co n sequence (e.g., history collection, sequence-of-events, preventive maintenance)

For that reason, a general process to establish a coexistence management for end user is described. In relation to the application classes defined it is recommended to assign priorities to the different wireless solutions. The intensity for the frequency management shall be correlated to the assigned priority classes. The process comprises the whole plant location and shall include all persons responsible for planning, installing, and commissioning of wireless devices. Wireless applications either in automation, logistic, or IT have to be considered. The coexistence management is a cyclic process which comprises all stages of stock taking, planning, installation, commissioning, maintenance, operation, and documentation of wireless applications at a location. It is further recommended to involve qualified service providers and own personnel at early phases, in avoidance of malfunctions in the long run.

Upcoming Wireless Base Technologies

The development of wireless technologies and extended standards is fast pacing. Especially the progress with respect to ultra wideband (UWB) represents a great potential, to open up new domains of applications in factory automation. First efforts for a standardisation of UWB technologies were initiated by the IEEE 802.15 WPAN High Rate Alternative PHY Task Group 3a (TG3a), founded in 2001. The task groups aim was to develop a high speed UWB technology, supporting data rates of >100Mbps at distances of <10m. Unfortunately, the group was not able to reach a consensus between two approaches offered by the leading industrial consortiums of the "WiMedia Alliance" and the "UWB Forum" and hence, disbanded in 2006. However, the approach of the WiMedia Alliance was published as the standard ECMA-368 in 2006 and is available in version 3.0 since 2008. The standard uses "Multiband OFDM" (MB-OFDM) as modulation and supports data rates of up to 480Mbps at distances of <10m. MB-OFDM is the basis of "Certified Wireless USB" (CW-USB). The application as an "Alternate MAC PHY" (AMP) is evaluated by the Bluetooth SIG. First transceiver chips and products are available since 2007. In 2007 the IEEE 802.15 WPAN Low Rate Alternative PHY Task Group 4a (TG4a) published the second UWB standard. IEEE 802.15.4a is a low data rate UWB technology supporting data rates of 0.1Mbps...27Mbps. It targets industrial sensor networks with real-time location capabilities. First transceiver chips will be available in 2010.

Ultra Wideband

In principle UWB is an old technology, whose origins come from military applications of the USA, more than 40 years ago. Whilst back then UWB was used as a tap-proof radio communication, nowadays the applications aim at high speed data transfers and real-time location systems. The first regulation for UWB devices, published by the FCC in 2002, defines UWB as follows. The relative bandwidth has to be larger than 20% and the absolute bandwidth has to be at least 500MHz at a 10dB cut-off frequency. The regulation gives no restrictions concerning signal forming and modulation. Because of the dense occupied frequency spectrum, UWB follows the approach of a parallel utilisation of the spectrum with a large bandwidth and a low spectral density power. In doing so, UWB appears as noise to coexisting narrow band technologies.

Classically, UWB is based on "Impulse Radio", which transmits information via impulses in the baseband without modulation. The UWB spectrum is generated due to extreme short durations (<1ns) of these impulses. Hence, UWB has the following inherent characteristics:

- Low latency times, due to extreme short symbol durations, what additionally offers the possibilities for precise ranging.
- Robust against the effects caused by multipath scattering. Reflexion and scattering are frequency selective. Using a high bandwidth reduces the probability of deep fading.
- Energy efficiency, due to the low spectral density power.
- Data rates of up to several Gbps.

Especially the first characteristics prove the potential of UWB for industrial communication systems. A typical use case would be a cable replacement for high speed real-time Ethernet field buses. Further use cases are WSANs.

Regulation for Ultra Wideband

Since UWB uses frequencies, which are already in use by licensed radio applications, the regulations are relatively restricted in order to avoid interferences to these applications. The first regulation was published by the FCC Part 15 Subpart F in 2002. The document defines seven classes of UWB devices. The classes of importance for factory automation are "Indoor UWB systems" and "Hand held UWB systems". "Indoor UWB systems" may only be used inside buildings and must have a fixed indoor infrastructure (i.e. power supply). "Hand held UWB systems" may operate indoor or outdoor and must not have a fixed infrastructure. The frequency ranges and maximal allowed transmit powers are depicted in table.

Table: Maximum EIRP spectral power densities for "Indoor UWB systems" and "Hand held systems" defined by FCC Part 15 Subpart F.

Frequency [MHz]	Max. EIRP [dBm/MHz]	
	Indoor UWB systems	Hand held UWB systems
960-1.610	-75.3	-75.3
1.610-1.990	-53.3	-63.3

1.990-3.100	-51. 3	-61. 3
3.100-10.600	-41. 3	-41. 3
<10.600	-51. 3	-61. 3

In further avoidance of interferences to GPS applications the maximal EIRP power density is limited to -83.3dBm/kHz for the frequency ranges of 1.164GHz...1.240GHz and 1.559MHz...1.610MHz. Within a frequency spectrum of 50MHz the maximal power is limited to 0dBm. It is obvious that the actual range of operation is between 3.1GHz...10.6GHz.

The regulation for the frequency range from 3.1GHz...10.6GHz for harmonised utilisation of UWB systems in Europe was released in 2007 by the decision of the European commission. The decision defines maximal EIRP power densities in dBm/MHz and within a spectrum of 50MHz. In addition to that, the decision differentiates between devices, which implement mitigation techniques in order to increase protection for radio Services. One mitigation technique is defined as "low duty cycle" (LDC). Devices implementing LDC must have a duty cycle lower than 0.5% per hour and lower than 5% per second. Furthermore, a single transmit duration must not exceed 5ms. Another technique is "detect and avoid" (DAA). Devices implementing DAA shall observe the used frequency spectrum with respect to coexisting devices and must adapt their transmit behaviour to avoid interferences.

Table: Maximum EIRP spectral power densities for Europe.

Frequency (GHz)	Max. EIRP Power Density (dBm/MHz)	Max. EIRP Power Density (dBm/50 MHz)
<1.6	-90. 0	-5 0. 0
1. 6-3.4	-85. 0	-4 5. 0
3.4-3. 8	-85. 0	- -45. 0
3.8-4. 2	-70. 0	- 30. 0
4.2-4. 8	-41. 3 (-70. 0)	-0.0 (-30,0) 1
4.8-6. 0	-70. 0	-30. 0
6.0-8. 5	-41. 3	-0.0
8. 5-10. 6	-65. 0	- 25. 0
"/10. 6	-85. 0	-45. 0

Table above shows, that the utilisation of frequencies below 6GHz will be restricted to devices, implementing mitigation techniques. How far real-time applications in factory automation can be served, regarding these restrictions, has to be investigated.

Industrial Internet of Things

The Industrial Internet of Things (IIoT) refers to interconnected sensors, instruments, and other devices networked together with computers' industrial applications, including manufacturing and energy management. This connectivity allows for data collection, exchange, and analysis, potentially facilitating improvements in productivity and efficiency as well as other economic benefits. The IIoT is an evolution of a distributed control system (DCS) that allows for a higher degree of automation by using cloud computing to refine and optimize the process controls.

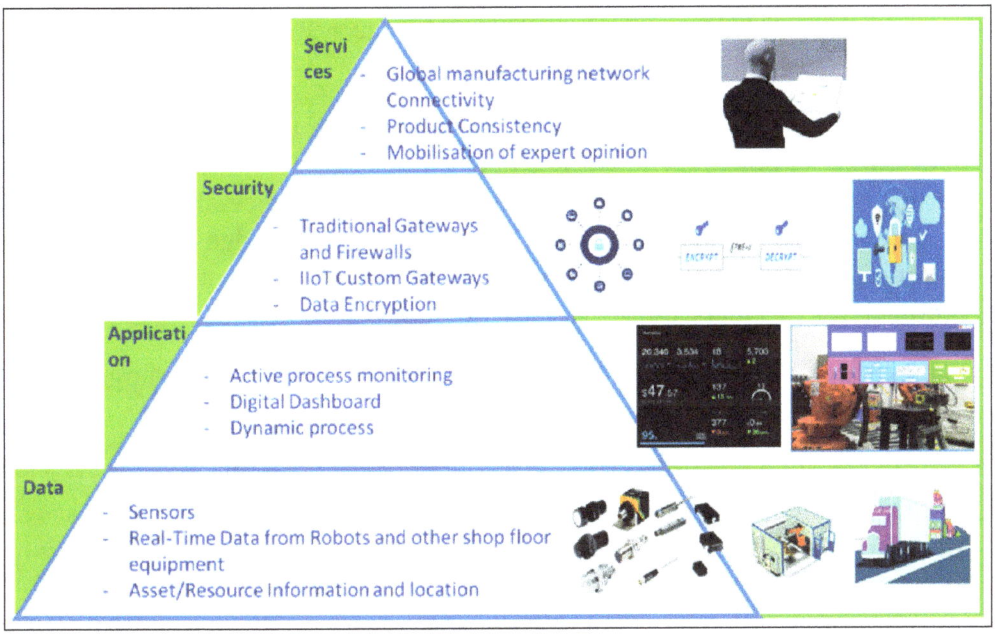

IIoT Architecture.

IIoT.

The IIoT is enabled by technologies such as cybersecurity, cloud computing, edge computing, mobile technologies, machine-to-machine, 3D printing, advanced robotics, big data, internet of things, RFID technology, and cognitive computing. Five of the most important ones are described below:

- Cyber-physical systems (CPS): The basic technology platform for IoT and IIoT and therefore the main enabler to connect physical machines that were previously disconnected. CPS integrates the dynamics of the physical process with those of software and communication, providing abstractions and modeling, design, and analysis techniques.

- Cloud computing: With cloud computing IT services and resources can be uploaded to and retrieved from the Internet as opposed to direct connection to a server. Files can be kept on cloud-based storage systems rather than on local storage devices.
- Edge computing: A distributed computing paradigm which brings computer data storage closer to the location where it is needed. In contrast to cloud computing, edge computing refers to decentralized data processing at the edge of the network. The industrial internet requires more of an edge-plus-cloud architecture rather than one based on purely centralized cloud; in order to transform productivity, products and services in the industrial world.
- Big data analytics: Big data analytics is the process of examining large and varied data sets, or big data.
- Artificial intelligence and machine learning: Artificial intelligence (AI) is a field within computer science in which intelligent machines are created that work and react like humans. Machine learning is a core part of AI, allowing software to more accurately predict outcomes without explicitly being programmed.

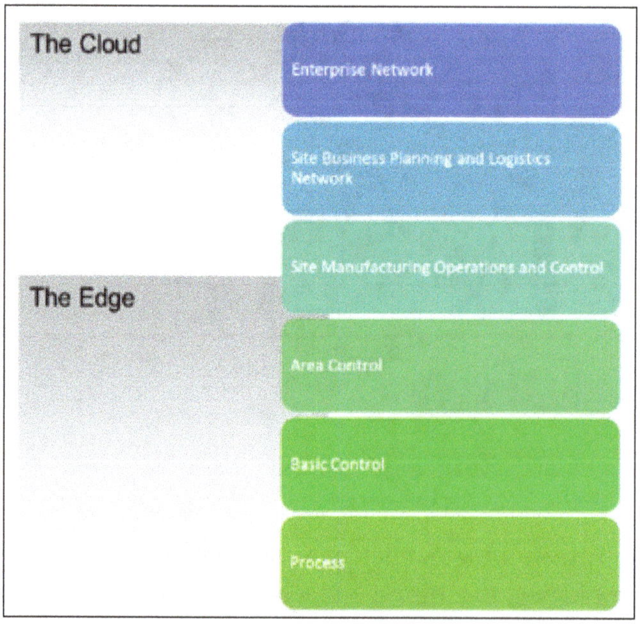

Approximate correspondence between levels in the Purdue model and the basic structure of the IoT.

IIoT systems are often conceived as a layered modular architecture of digital technology. The device layer refers to the physical components: CPS, sensors or machines. The network layer consists of physical network buses, cloud computing and communication protocols that aggregate and transport the data to the service layer, which consists of applications that manipulate and combine data into information that can be displayed on the driver dashboard. The top-most stratum of the stack is the content layer or the user interface.

Table: Layered modular architecture IIoT.

Content layer	User interface devices (e.g. screens, tablets, smart glasses).
Service layer	Applications, software to analyze data and transform it into information.
Network layer	Communications protocols, wifi, cloud computing.
Device layer	Hardware: CPS, machines, sensors.

The history of the IIoT begins with the invention of the programmable logic controller (PLC) by Dick Morley in 1968, which was used by General Motors in their automatic transmission manufacturing division. These PLCs allowed for fine control of individual elements in the manufacturing chain. In 1975, Honeywell and Yokogawa introduced the world's first DCSs, the TDC 2000 and the CENTUM system, respectively. These DCSs were the next step in allowing flexible process control throughout a plant, with the added benefit of backup redundancies by distributing control across the entire system, eliminating a singular point of failure in a central control room.

With the introduction of Ethernet in 1980, people began to explore the concept of a network of smart devices as early as 1982, when a modified Coke machine at Carnegie Mellon University became the first internet-connected appliance, able to report its inventory and whether newly loaded drinks were cold. As early as in 1994, greater industrial applications were envisioned, as Reza Raji described the concept in IEEE Spectrum as "moving small packets of data to a large set of nodes, so as to integrate and automate everything from home appliances to entire factories".

The concept of the internet of things first became popular in 1999, through the Auto-ID Center at MIT and related market-analysis publications. Radio-frequency identification (RFID) was seen by Kevin Ashton (one of the founders of the original Auto-ID Center) as a prerequisite for the internet of things at that point. If all objects and people in daily life were equipped with identifiers, computers could manage and inventory them. Besides using RFID, the tagging of things may be achieved through such technologies as near field communication, barcodes, QR codes and digital watermarking.

The current conception of the IIoT arose after the emergence of cloud technology in 2002, which allows for the storage of data to examine for historical trends, and the development of the OPC Unified Architecture protocol in 2006, which enabled secure, remote communications between devices, programs, and data sources without the need for human intervention or interfaces.

One of the first consequences of implementing the industrial internet of things (by equipping objects with minuscule identifying devices or machine-readable identifiers) would be to create instant and ceaseless inventory control. Another benefit of implementing an IIoT system is the ability to create a digital twin of the system. Utilizing this digital twin allows for further optimization of the system by allowing for experimentation with new data from the cloud without having to halt production or sacrifice safety, as the new processes can be refined virtually until they are ready to be implemented. A digital twin can also serve as a training ground for new employees who won't have to worry about real impacts on the live system.

Standards and Frameworks

IoT frameworks help support the interaction between "things" and allow for more complex structures like distributed computing and the development of distributed applications. Currently, some IoT frameworks focus on real-time data logging solutions like Jasper Technologies, Inc. and Xively: offering some basis to work with many "things" and have them interact. Future developments may lead to software development environments targeted specifically for creating the software needed to work with IoT hardware. Companies are developing technology platforms to provide this type of functionality for the internet of things. Newer platforms are being developed, which add more intelligence.

- IBM has announced cognitive IoT, which combines traditional IoT with machine intelligence and learning, contextual information, industry-specific models and natural language processing.
- The XMPP Standards Foundation (XSF) is creating such a framework called Chatty Things, which is a fully open, vendor-independent standard using XMPP to provide a distributed, scalable, and secure infrastructure.
- REST is a scalable architecture which allows for things to communicate over Hypertext Transfer Protocol and is easily adopted for IoT applications to provide communication from a thing to a central web server.
- MQTT is a publish-subscribe architecture on top of TCP/IP which allows for bi-directional communication between a thing and a MQTT broker.
- Node-RED in an open source software designed by IBM to connect APIs, hardware, and online services.
- OPC is a series of standards designed by the OPC Foundation to connect computer systems to automated devices.
- The Industrial Internet Consortium's (IIC) Industrial Internet Reference Architecture (IIRA) and the German Industry 4.0 are independent efforts to create a defined standard for IIoT-enabled facilities.

Application and Industries

The term industrial internet of things is often encountered in the manufacturing industries, referring to the industrial subset of the IoT. The industrial internet of things will enable the creation of new business models by improving productivity, exploiting analytics for innovation, and transforming the workforce. The potential of growth by implementing IIoT is predicted to generate $15 trillion of global GDP by 2030.

While connectivity and data acquisition are imperative for IIoT, they are not the end goals, but rather the foundation and path to something bigger. Of all the technologies, predictive maintenance is an "easier" application, as it is applicable to existing assets and management systems. Intelligent maintenance systems can reduce unexpected downtime and increase productivity, which is projected to save up to 12% over scheduled repairs, reduce overall maintenance costs up to 30%, and eliminate breakdowns up to 70%, according to some studies. Industrial big data analytics plays a vital role in manufacturing asset predictive maintenance, although that is not the only capability of industrial big data. Cyber-physical systems (CPS) are the core technology of industrial big data and they will be an interface between human and the cyber world. Cyber-physical systems can be designed by following the 5C (connection, conversion, cyber, cognition, configuration) architecture, and they transform the collected data into actionable information and eventually interact with the physical assets to optimize processes.

An IoT-enabled intelligent system of such capability has been demonstrated by the NSF Industry/University Collaborative Research Center for Intelligent Maintenance Systems (IMS) at University of Cincinnati on a band saw machine in IMTS 2014 in Chicago. Band saw machines are not necessarily expensive, but band saw belt expenses are enormous since they degrade much faster. However, without sensing and intelligent analytics, it can only be determined by experience when the band saw belt will actually break. The developed prognostics system is able to recognize and monitor the degradation of band saw belts even if the condition is changing, so that users can know in near real-time the optimal time to replace the belt. The developed analytical algorithms were realized on a cloud server, and were made accessible via the Internet and on mobile devices.

Integration of sensing and actuation systems connected to the Internet can optimize energy consumption as a whole. It is expected that IoT devices will be integrated into all forms of energy consuming devices (switches, power outlets, bulbs, televisions, etc.) and be able to communicate with the utility supply company in order to effectively balance power generation and energy usage. Besides home based energy management, the IIoT is especially relevant to the Smart Grid since it provides systems to gather and act on energy and power-related information in an automated fashion with the goal to improve the efficiency, reliability, economics, and sustainability of the production and distribution of electricity. Using advanced metering infrastructure (AMI) devices connected to the Internet backbone, electric utilities can not only collect data from end-user connections, but also manage other distribution automation devices like transformers and reclosers.

As of 2016, other real-world applications include incorporating smart LEDs to direct shoppers to empty parking spaces or highlight shifting traffic patterns, using of sensors on water purifiers to alert managers via computer or smartphone when to replace parts, attaching RFID tags to safety gear to track personnel and ensure their safety, embedding computers into power tools to record and track the torque level of individual tightenings, and collecting data from multiple systems to enable the simulation of new processes.

Automotive Industry

Using IIoT in car manufacturing implies the digitalization of all elements of production. Software, machines and humans are interconnected, enabling suppliers and manufacturers to rapidly respond to changing standards. IIoT enables efficient and cost-effective production by moving data from the customers to the company's systems, and then to individual sections of the production process. With IIoT, new tools and functionalities can be included in the manufacturing process. For example, 3D printers simplify the way of shaping pressing tools by printing the shape directly from steel granulate. These tools enable new possibilities for designing (with high precision). Customization of vehicles is also enabled by IIoT due to the modularity and connectivity of this technology. While in the past they worked separately, IIoT now enables humans and robots to cooperate. Robots take on the heavy and repetitive activities, so the manufacturing cycles are quicker and the vehicle comes to the market more rapidly. Factories can quickly identify potential maintenance issues before they lead to downtime and many of them are moving to a 24-hour production plant, due to higher security and efficiency. The majority of automotive manufacturers companies have production plants in different countries, where different components of the same vehicle are built. IIoT makes possible to connect these production plants to each other, creating the possibility to move within facilities. Big data can be visually monitored which enables companies to respond faster to fluctuations in production and demand.

Oil and Gas Industry

With IIoT support, large amounts of raw data can be stored and sent by the drilling gear and research stations for cloud storage and analysis. With IIoT technologies, the oil and gas industry has the capability to connect machines, devices, sensors, and people through interconnectivity, which can help companies better address fluctuations in demand and pricing, address cybersecurity, and minimize environmental impact.

Across the supply chain, IIoT can improve the maintenance process, the overall safety, and connectivity. Drones can be used to detect possible oil and gas leaks at an early stage and at locations that are difficult to reach (e.g. offshore). They can also be used to identify weak spots in complex networks of pipelines with built-in thermal imaging systems. Increased connectivity (data integration and communication) can help companies with adjusting the production levels based on real-time data of inventory, storage, distribution pace, and forecasted demand. For example, a Deloitte report states that by implementing an IIoT solution integrating data from multiple internal and external sources (such as work management system, control centre, pipeline attributes, risk scores, inline inspection findings, planned assessments, and leak history), thousands of miles of pipes can be monitored in real-time. This allows monitoring of pipeline threats, improving risk management, and providing situational awareness.

Benefits also apply to specific processes of the oil and gas industry. The exploration process of oil and gas can be done more precisely with 4D models built by seismic imaging. These models map fluctuations in oil reserves and gas levels, they strive to point out the exact quantity of resources needed, and they forecast the lifespan of wells. The application of smart sensors and automated drillers gives companies the opportunity to monitor and produce more efficiently. Further, the storing process can also be improved with the implementation of IIoT by collecting and analyzing real-time data to monitor inventory levels and temperature control. IIoT can enhance the transportation process of oil and gas by implementing smart sensors and thermal detectors to give real-time geolocation data and monitor the products for safety reasons. These smart sensors can monitor the refinery processes, and enhance safety. The demand for products can be forecasted more precisely and automatically be communicated to the refineries and production plants to adjust production levels.

Security

As the IIoT expands, new security concerns arise with it. Every new device or component that connects to the IIoT can become a potential liability. Gartner estimates that by 2020, more than 25% of recognized attacks on enterprises will involve IoT-connected systems, despite its accounting for less than 10% of IT security budgets. Existing cybersecurity measures are vastly inferior for internet-connected devices compared to their traditional computer counterparts, which can allow for them to be hijacked for DDoS-based attacks by botnets like Mirai. Another possibility is the infection of internet-connected industrial controllers, like in the case of Stuxnet, without the need for physical access to the system to spread the worm.

Additionally, IIoT-enabled devices can allow for more "traditional" forms of cybercrime, as in the case of the 2013 Target data breach, where information was stolen after hackers gained access to Target's networks via credentials stolen from a third party HVAC vendor. The pharmaceutical manufacturing industry has been slow to adopt IIoT advances because of security concerns such as these. One of the difficulties in providing security solutions in IIoT applications is the fragmented nature of the hardware. Consequently, security architectures are turning towards designs that are software-based or device-agnostic.

Industrial Control System

Industrial control system (ICS) is a general term that encompasses several types of control systems and associated instrumentation used for industrial process control.

Such systems can range from a few modular panel-mounted controllers to large interconnected and interactive distributed control systems with many thousands of field connections. All systems receive data received from remote sensors measuring process variables (PVs), compare these with desired set points (SPs) and derive command functions which are used to control a process through the final control elements (FCEs), such as control valves.

The larger systems are usually implemented by Supervisory Control and Data Acquisition (SCADA) systems, or distributed control systems (DCS), and programmable logic controllers (PLCs), though SCADA and PLC systems are scalable down to small systems with few control loops. Such systems are extensively used in industries such as chemical processing, pulp and paper manufacture, power generation, oil and gas processing and telecommunications.

Discrete Controllers

Panel mounted controllers with integral displays. The process value (PV), and setvalue (SV) or setpoint are on the same scale for easy comparison. The controller output is shown as MV (manipulated variable) with range 0-100%.

The simplest control systems are based around small discrete controllers with a single control loop each. These are usually panel mounted which allows direct viewing of the front panel and provides means of manual intervention by the operator, either to manually control the process or to change control setpoints. Originally these would be pneumatic controllers, a few of which are still in use, but nearly all are now electronic.

Quite complex systems can be created with networks of these controllers communicating using industry standard protocols. Networking allow the use of local or remote SCADA operator interfaces, and enables the cascading and interlocking of controllers. However, as the number of control loops increase for a system design there is a point where the use of a programmable logic controller (PLC) or distributed control system (DCS) is more manageable or cost-effective.

A distributed control system (DCS) is a digital processor control system for a process or plant, wherein controller functions and field connection modules are distributed throughout the system. As the number of control loops grows, DCS becomes more cost effective than discrete controllers. Additionally a DCS provides supervisory viewing and management over large industrial processes. In a DCS, a hierarchy of controllers is connected by communication networks, allowing centralised control rooms and local on-plant monitoring and control.

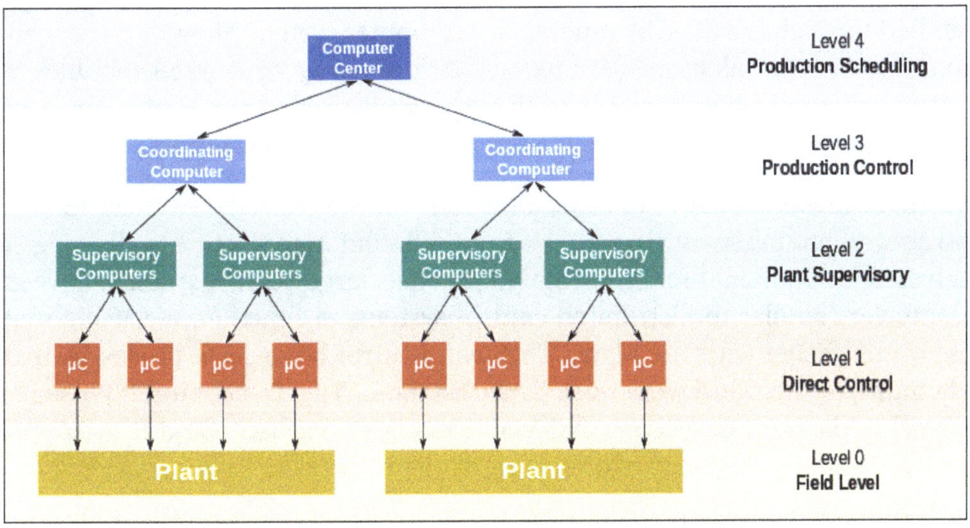

Functional manufacturing control levels: DCS (including PLCs or RTUs) operate on level 1. Level 2 contains the SCADA software and computing platform.

A DCS enables easy configuration of plant controls such as cascaded loops and interlocks, and easy interfacing with other computer systems such as production control. It also enables more sophisticated alarm handling, introduces automatic event logging, removes the need for physical records such as chart recorders and allows the control equipment to be networked and thereby located locally to equipment being controlled to reduce cabling. A DCS typically uses custom-designed processors as controllers, and uses either proprietary interconnections or standard protocols for communication. Input and output modules form the peripheral components of the system.

The processors receive information from input modules, process the information and decide control actions to be performed by the output modules. The input modules receive information from sensing instruments in the process (or field) and the output modules transmit instructions to the final control elements, such as control valves. The field inputs and outputs can either be continuously changing analog signals e.g. current loop or 2 state signals that switch either *on* or *off*, such as relay contacts or a semiconductor switch.

Distributed control systems can normally also support Foundation Fieldbus, PROFIBUS, HART, Modbus and other digital communication buses that carry not only input and output signals but also advanced messages such as error diagnostics and status signals.

SCADA Systems

Supervisory control and data acquisition (SCADA) is a control system architecture that uses computers, networked data communications and graphical user interfaces for high-level process supervisory management. The operator interfaces which enable monitoring and the issuing of process commands, such as controller set point changes, are handled through the SCADA supervisory computer system. However, the real-time control logic or controller calculations are performed by networked modules which connect to other peripheral devices such as programmable logic controllers and discrete PID controllers which interface to the process plant or machinery.

The SCADA concept was developed as a universal means of remote access to a variety of local control modules, which could be from different manufacturers allowing access through standard automation protocols. In practice, large SCADA systems have grown to become very similar to distributed control systems in function, but using multiple means of interfacing with the plant. They can control large-scale processes that can include multiple sites, and work over large distances. This is a commonly-used architecture industrial control systems, however there are concerns about SCADA systems being vulnerable to cyberwarfare or cyberterrorism attacks.

The SCADA software operates on a supervisory level as control actions are performed automatically by RTUs or PLCs. SCADA control functions are usually restricted to basic overriding or supervisory level intervention. A feedback control loop is directly controlled by the RTU or PLC, but the SCADA software monitors the overall performance of the loop. For example, a PLC may control the flow of cooling water through part of an industrial process to a set point level, but the SCADA system software will allow operators to change the set points for the flow. The SCADA also enables alarm conditions, such as loss of flow or high temperature, to be displayed and recorded.

Programmable Logic Controllers

PLCs can range from small modular devices with tens of inputs and outputs (I/O) in a housing integral with the processor, to large rack-mounted modular devices with a count of thousands of I/O, and which are often networked to other PLC and SCADA systems. They can be designed for multiple arrangements of digital and analog inputs and outputs, extended temperature ranges, immunity to electrical noise, and resistance to vibration and impact. Programs to control machine operation are typically stored in battery-backed-up or non-volatile memory.

Siemens Simatic S7-400 system in a rack, left-to-right: power supply unit (PSU), CPU, interface module (IM) and communication processor (CP).

Industrial Robots

An industrial robot is a robot system used for manufacturing. Industrial robots are automated, programmable and capable of movement on three or more axis.

Articulated industrial robot operating in a foundry.

Typical applications of robots include welding, painting, assembly, disassembly, pick and place for printed circuit boards, packaging and labeling, palletizing, product inspection, and testing; all accomplished with high endurance, speed, and precision. They can assist in material handling.

Types and Features

A set of six-axis robots used for welding.

The most commonly used robot configurations are articulated robots, SCARA robots, delta robots and cartesian coordinate robots, (gantry robots or x-y-z robots). In the context of general robotics, most types of robots would fall into the category of robotic arms (inherent in the use of the word manipulator in ISO standard 8373). Robots exhibit varying degrees of autonomy:

- Some robots are programmed to faithfully carry out specific actions over and over again (repetitive actions) without variation and with a high degree of accuracy. These actions are determined by programmed routines that specify the direction, acceleration, velocity, deceleration, and distance of a series of coordinated motions.
- Other robots are much more flexible as to the orientation of the object on which they are operating or even the task that has to be performed on the object itself, which the robot may even need to identify. For example, for more precise guidance, robots often contain machine vision sub-systems acting as their visual sensors, linked to powerful computers or controllers. Artificial intelligence, or what passes for it, is becoming an increasingly important factor in the modern industrial robot.

Technical Description

Defining Parameters

- Number of axes: Two axes are required to reach any point in a plane; three axes are required to reach any point in space. To fully control the orientation of the end of the arm(i.e. the *wrist*) three more axes (yaw, pitch, and roll) are required. Some designs (e.g. the SCARA robot) trade limitations in motion possibilities for cost, speed, and accuracy.
- Degrees of freedom: This is usually the same as the number of axes.
- Working envelope: The region of space a robot can reach.
- Kinematics: The actual arrangement of rigid members and joints in the robot, which determines the robot's possible motions. Classes of robot kinematics include articulated, cartesian, parallel and SCARA.

- Carrying capacity or payload: How much weight a robot can lift.
- Speed: How fast the robot can position the end of its arm. This may be defined in terms of the angular or linear speed of each axis or as a compound speed i.e. the speed of the end of the arm when all axes are moving.
- Acceleration: How quickly an axis can accelerate. Since this is a limiting factor a robot may not be able to reach its specified maximum speed for movements over a short distance or a complex path requiring frequent changes of direction.
- Accuracy: How closely a robot can reach a commanded position. When the absolute position of the robot is measured and compared to the commanded position the error is a measure of accuracy. Accuracy can be improved with external sensing for example a vision system or Infra-Red. Accuracy can vary with speed and position within the working envelope and with payload.
- Repeatability: How well the robot will return to a programmed position. This is not the same as accuracy. It may be that when told to go to a certain X-Y-Z position that it gets only to within 1 mm of that position. This would be its accuracy which may be improved by calibration. But if that position is taught into controller memory and each time it is sent there it returns to within 0.1mm of the taught position then the repeatability will be within 0.1mm.

Accuracy and repeatability are different measures. Repeatability is usually the most important criterion for a robot and is similar to the concept of 'precision' in measurement. ISO 9283 sets out a method whereby both accuracy and repeatability can be measured. Typically a robot is sent to a taught position a number of times and the error is measured at each return to the position after visiting 4 other positions. Repeatability is then quantified using the standard deviation of those samples in all three dimensions. A typical robot can, of course make a positional error exceeding that and that could be a problem for the process. Moreover, the repeatability is different in different parts of the working envelope and also changes with speed and payload. ISO 9283 specifies that accuracy and repeatability should be measured at maximum speed and at maximum payload. But this results in pessimistic values whereas the robot could be much more accurate and repeatable at light loads and speeds. Repeatability in an industrial process is also subject to the accuracy of the end effector, for example a gripper, and even to the design of the 'fingers' that match the gripper to the object being grasped. For example, if a robot picks a screw by its head, the screw could be at a random angle. A subsequent attempt to insert the screw into a hole could easily fail. These and similar scenarios can be improved with 'lead-ins' e.g. by making the entrance to the hole tapered.

- Motion control: For some applications, such as simple pick-and-place assembly, the robot need merely return repeatably to a limited number of pre-taught positions. For more sophisticated applications, such as welding and finishing (spray painting), motion must be continuously controlled to follow a path in space, with controlled orientation and velocity.

- Power source: Some robots use electric motors, others use hydraulic actuators. The former are faster, the latter are stronger and advantageous in applications such as spray painting, where a spark could set off an explosion; however, low internal air-pressurisation of the arm can prevent ingress of flammable vapours as well as other contaminants. Nowadays, it is highly unlikely to see any hydraulic robots in the market. Additional sealings, brushless electric motors and spark-proof protection eased the construction of units that are able to work in the environment with an explosive atmosphere.
- Drive: Some robots connect electric motors to the joints via gears; others connect the motor to the joint directly (direct drive). Using gears results in measurable 'backlash' which is free movement in an axis. Smaller robot arms frequently employ high speed, low torque DC motors, which generally require high gearing ratios; this has the disadvantage of backlash. In such cases the harmonic drive is often used.
- Compliance: This is a measure of the amount in angle or distance that a robot axis will move when a force is applied to it. Because of compliance when a robot goes to a position carrying its maximum payload it will be at a position slightly lower than when it is carrying no payload. Compliance can also be responsible for overshoot when carrying high payloads in which case acceleration would need to be reduced.

Robot Programming and Interfaces

The setup or programming of motions and sequences for an industrial robot is typically taught by linking the robot controller to a laptop, desktop computer or (internal or Internet) network.

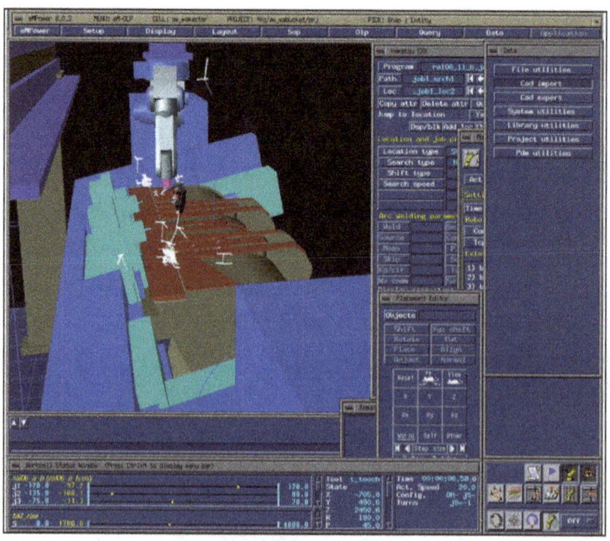

Offline programming.

A robot and a collection of machines or peripherals is referred to as a workcell, or cell. A typical cell might contain a parts feeder, a molding machine and a robot. The various machines are 'integrated' and controlled by a single computer or PLC. How the robot interacts with other machines in the cell must be programmed, both with regard to their positions in the cell and synchronizing with them.

Software

The computer is installed with corresponding interface software. The use of a computer greatly simplifies the programming process. Specialized robot software is run either in the robot controller or in the computer or both depending on the system design.

There are two basic entities that need to be taught (or programmed): positional data and procedure. For example, in a task to move a screw from a feeder to a hole the positions of the feeder and the hole must first be taught or programmed. Secondly the procedure to get the screw from the feeder to the hole must be programmed along with any I/O involved, for example a signal to indicate when the screw is in the feeder ready to be picked up. The purpose of the robot software is to facilitate both these programming tasks.

Teaching the robot positions may be achieved a number of ways:

- Positional commands: The robot can be directed to the required position using a GUI or text based commands in which the required X-Y-Z position may be specified and edited.

A typical well-used teach pendant with optional mouse.

- Teach pendant*:* Robot positions can be taught via a teach pendant. This is a handheld control and programming unit. The common features of such units are the ability to manually send the robot to a desired position, or "inch" or "jog" to adjust a position. They also have a means to change the speed since a low speed is usually required for careful positioning, or while test-running through a new or modified routine. A large emergency stop button is usually included as well. Typically once the robot has been programmed there is no more use for the teach pendant. All teach pendants are equipped with a 3-position deadman switch. In the manual mode, it allows the robot to move only when it is in the middle position (partially pressed). If it is fully pressed in or completely released, the robot stops. This principle of operation allows natural reflexes to be used to increase safety.

- Lead-by-the-nose: This is a technique offered by many robot manufacturers. In this method, one user holds the robot's manipulator, while another person enters a command which de-energizes the robot causing it to go into limp. The user then moves the robot by hand to the required positions and along a required path while the software logs these positions into memory. The program can later run the robot to these positions or along the taught path. This technique is popular for tasks such as paint spraying.
- Offline programming is where the entire cell, the robot and all the machines or instruments in the workspace are mapped graphically. The robot can then be moved on screen and the process simulated. A robotics simulator is used to create embedded applications for a robot, without depending on the physical operation of the robot arm and end effector. The advantages of robotics simulation is that it saves time in the design of robotics applications. It can also increase the level of safety associated with robotic equipment since various "what if" scenarios can be tried and tested before the system is activated. Robot simulation software provides a platform to teach, test, run, and debug programs that have been written in a variety of programming languages.

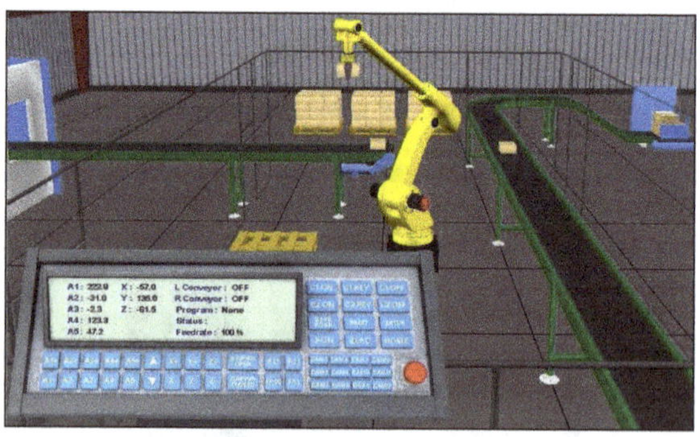

Robotics Simulator.

- Robot simulation tools allow for robotics programs to be conveniently written and debugged off-line with the final version of the program tested on an actual robot. The ability to preview the behavior of a robotic system in a virtual world allows for a variety of mechanisms, devices, configurations and controllers to be tried and tested before being applied to a "real world" system. Robotics simulators have the ability to provide real-time computing of the simulated motion of an industrial robot using both geometric modeling and kinematics modeling.
- Others in addition, machine operators often use user interface devices, typically touchscreen units, which serve as the operator control panel. The operator

can switch from program to program, make adjustments within a program and also operate a host of peripheral devices that may be integrated within the same robotic system. These include end effectors, feeders that supply components to the robot, conveyor belts, emergency stop controls, machine vision systems, safety interlock systems, barcode printers and an almost infinite array of other industrial devices which are accessed and controlled via the operator control panel.

The teach pendant or PC is usually disconnected after programming and the robot then runs on the program that has been installed in its controller. However a computer is often used to 'supervise' the robot and any peripherals, or to provide additional storage for access to numerous complex paths and routines.

End-of-arm Tooling

The most essential robot peripheral is the end effector, or end-of-arm-tooling (EOT). Common examples of end effectors include welding devices (such as MIG-welding guns, spot-welders, etc.), spray guns and also grinding and deburring devices (such as pneumatic disk or belt grinders, burrs, etc.), and grippers (devices that can grasp an object, usually electromechanical or pneumatic). Other common means of picking up objects is by vacuum or magnets. End effectors are frequently highly complex, made to match the handled product and often capable of picking up an array of products at one time. They may utilize various sensors to aid the robot system in locating, handling, and positioning products.

Controlling Movement

For a given robot the only parameters necessary to completely locate the end effector (gripper, welding torch, etc.) of the robot are the angles of each of the joints or displacements of the linear axes (or combinations of the two for robot formats such as SCARA). However, there are many different ways to define the points. The most common and most convenient way of defining a point is to specify a Cartesian coordinate for it, i.e. the position of the 'end effector' in mm in the X, Y and Z directions relative to the robot's origin. In addition, depending on the types of joints a particular robot may have, the orientation of the end effector in yaw, pitch, and roll and the location of the tool point relative to the robot's faceplate must also be specified. For a jointed arm these coordinates must be converted to joint angles by the robot controller and such conversions are known as Cartesian Transformations which may need to be performed iteratively or recursively for a multiple axis robot. The mathematics of the relationship between joint angles and actual spatial coordinates is called kinematics.

Positioning by Cartesian coordinates may be done by entering the coordinates into the system or by using a teach pendant which moves the robot in X-Y-Z directions. It is much easier for a human operator to visualize motions up/down, left/right, etc. than to move each joint one at a time. When the desired position is reached it is then defined in some way particular to the robot software in use, e.g. P1 - P5 below.

Typical Programming

Most articulated robots perform by storing a series of positions in memory, and moving to them at various times in their programming sequence. For example, a robot which is moving items from one place to another might have a simple 'pick and place' program similar to the following:

Define points P1–P5:

- Safely above workpiece (defined as P1).
- 10 cm Above bin A (defined as P2).
- At position to take part from bin A (defined as P3).
- 10 cm Above bin B (defined as P4).
- At position to take part from bin B. (defined as P5).

Define program:

- Move to P1.
- Move to P2.
- Move to P3.
- Close gripper.
- Move to P2.
- Move to P4.
- Move to P5.
- Open gripper.
- Move to P4.
- Move to P1 and finish.

Singularities

The American National Standard for Industrial Robots and Robot Systems — Safety Requirements (ANSI/RIA R15.06-1999) defines a singularity as "a condition caused by the collinear alignment of two or more robot axes resulting in unpredictable robot motion and velocities." It is most common in robot arms that utilize a "triple-roll wrist". This is a wrist about which the three axes of the wrist, controlling yaw, pitch, and roll, all pass through a common point. An example of a wrist singularity is when the path through which the robot is traveling causes the first and third axes of the robot's wrist (i.e. robot's axes 4 and 6) to line up. The second wrist axis then attempts to spin 180° in zero time to maintain the orientation of the end effector. Another common term for this singularity is a "wrist flip". The result of a singularity can be quite dramatic and can have adverse effects on the robot arm, the end effector, and the process. Some industrial robot manufacturers have attempted to side-step the situation by slightly altering the robot's path to prevent this condition. Another method is to slow the robot's travel speed, thus reducing the speed required for the wrist to make the transition. The ANSI/RIA has mandated that robot manufacturers shall make the user aware of singularities if they occur while the system is being manually manipulated.

A second type of singularity in wrist-partitioned vertically articulated six-axis robots occurs when the wrist center lies on a cylinder that is centered about axis 1 and with radius equal to the distance between axes 1 and 4. This is called a shoulder singularity. Some robot manufacturers also mention alignment singularities, where axes 1 and 6 become coincident. This is simply a sub-case of shoulder singularities. When the robot passes close to a shoulder singularity, joint 1 spins very fast.

The third and last type of singularity in wrist-partitioned vertically articulated six-axis robots occurs when the wrist's center lies in the same plane as axes 2 and 3.

Singularities are closely related to the phenomena of gimbal lock, which has a similar root cause of axes becoming lined up.

Health and Safety

The International Federation of Robotics has predicted a worldwide increase in adoption of industrial robots and they estimated 1.7 million new robot installations in factories worldwide by 2020. Rapid advances in automation technologies (e.g. fixed robots, collaborative and mobile robots, and exoskeletons) have the potential to improve work conditions but also to introduce workplace hazards in manufacturing workplaces. Despite the lack of occupational surveillance data on injuries associated specifically with robots, researchers from the US National Institute for Occupational Safety and Health (NIOSH) identified 61 robot-related deaths between 1992 and 2015 using keyword searches of the Bureau of Labor Statistics (BLS) Census of Fatal Occupational Injuries research database. Using data from the Bureau of Labor Statistics, NIOSH and its state partners have investigated 4 robot-related fatalities under the Fatality Assessment and Control Evaluation Program. In addition the Occupational Safety and Health Administration (OSHA) has investigated dozens of robot-related deaths and injuries, which can be reviewed at OSHA Accident Search page. Injuries and fatalities could increase over time because of the increasing number of collaborative and co-existing robots, powered exoskeletons, and autonomous vehicles into the work environment.

Safety standards are being developed by the Robotic Industries Association (RIA) in conjunction with the American National Standards Institute (ANSI). On October 5, 2017, OSHA, NIOSH and RIA signed an alliance to work together to enhance technical expertise, identify and help address potential workplace hazards associated with traditional industrial robots and the emerging technology of human-robot collaboration installations and systems, and help identify needed research to reduce workplace hazards. On October 16 NIOSH launched the Center for Occupational Robotics Research to "provide scientific leadership to guide the development and use of occupational robots that enhance worker safety, health, and wellbeing." So far, the research needs identified by NIOSH and its partners include: tracking and preventing injuries and fatalities, intervention and dissemination strategies to promote safe machine control and maintenance procedures, and on translating effective evidence-based interventions into workplace practice.

Robotics industrial automation is changing the face of production. Manufacturers around the globe are implementing some form of automation to become more efficient, safe and ultimately to increase revenues.

Advantages of Industrial Automation

- Quality Control: No one likes purchasing an item, only to be disappointed by the poor quality. Robots are a great solution for higher quality production. Quality builds trust from customers as well as pride knowing that you are contributing something of value.
- Repeatability: Being consistent and knowing that you will get the same quality end product is critical to efficiency. A robot is able to perform the exact same task, exactly the same way, over and over again. Less errors means less wasted time.
- Waste Reduction: Consistent repeatability allows manufacturers to reduce overall waste. Less errors not only saves time, but it also reduces the amount of material required to produce the product. A couple of examples: robots can use less wire for welding, less amount of paint, and cut closer to the edge.
- Faster Cycle Times: Unfortunately humans have their limitations. Robots have been known to greatly improve production cycle speeds. The more you can produce, the higher demand you can meet and ultimately bringing in more money.
- Improved Workplace Safety: There are so many dangerous work environments that can have horrible side effects on a human body. Separating workers from lifting too much weight, exposure to fumes and gases, close interaction with lasers or blades, can tremendously decrease the possibility of injury.
- Reduction of Labor Costs: Labor can be expensive, especially when you factor in medical benefits, paid time off, injury comp time, etc. Robots can replace

certain jobs, but that doesn't mean they are going to take over the world. It just means we need to adjust our focus. Understanding that robots are there for our own safety and efficiency allows us to remove workers from tough tedious jobs, to more fulfilling roles.

- Reduced Floor Space: It's easy to start sprawling out across the shop floor with extra materials, tools, and machinery. Robots can help reduce the footprint of the required workspace by optimizing everything into a smaller, confined space.
- Integration with Business Systems: Now a days, communication between multiple data platforms is rapidly growing, improving efficiency. You can see when there's a bottleneck a lot quicker with proper technology installed. Robots and machinery are talking with one another to give business leaders a better view on the overall picture, helping them make smarter decisions on how to improve their process.

Engineering Processes for Decentralized Factory Automation Systems

The necessity to improve the current automation concepts for cost reduction in factory automation represents a widely discussed problem. Research has developed solutions for this problem area for quite some time based on decentralized automation concepts (e.g., holonic systems, agent systems). However, industrial companies have shown reluctance concerning a broad application in practice. The reasons reside mainly in lacking engineering methods for systematic implementation in industrial businesses and lacking reliable evaluation of the consequences for application domains over the entire life cycle.

The main goal for the development of decentralized automation concepts and systems is to achieve flexibility in factory automation systems, driven by ever more rapidly changing production conditions, such as order variations, changing products, load variations, or plug & produce capabilities of machines. Hence, the value promised by decentralized automation concepts mainly resides in the improvement of operative parameters of a plant, for instance, through more flexibility of usage, higher efficiency, or availability. Such parameters were evaluated based on prototypes, as well as by means of simulation, and were verified with a relatively good validity.

However, the costs for introducing and applying decentralized automation concepts and systems – in terms of total cost of ownership (TCO) – have not been properly investigated yet, neither in science nor in industrial application. The reason is that we know much about the operation phase behaviour of production facilities using decentralized automation systems, but there exists hardly any explanation of the impact, in terms of benefits and risks, on the entire industrial life cycle. In addition to operating costs, the second most important aspect of TCO lies in the activities and processes for engineering (design, realization, and commissioning) of production facilities. The related cost potential is certainly significant. In 2005, automotive manufacturers identified the portion of the cost for engineering tasks in automotive manufacturing to be 20 to 25% of the total investment for the production line and 55% for the control system – with an upward trend. Hence, the requirements for engineering of such facilities must be considered. The main objective here is to identify the impact of automation concepts on engineering activities, to create suitable and effective engineering methods for decentralized automation, and to render them applicable for industrial applications.

Basic Approach of (Intelligent) Decentralized Automation

The factory automation is structured hierarchically in form of an "automation pyramid," in which each level performs special automation functions. Individual automation devices/systems, such as field devices, PLC, HMI, and control systems, are specifically attributed to individual levels. Data generated in industrial production are strictly attributed to the different levels, transferred via interfaces, and aggregated or abstracted.

Table: Automation levels and systems.

Label	Typical Systems	Tasks (excerpt)
Enterprise level	Enterprise Resource Planning Systems	Rough production planning Order processing, logistics.
Production level	Manufacturing Execution, or Warehouse Management Systems	Order management: Detailed planning, control, and monitoring of production or material flow Management of plant information, reports, alarms Management of resources, personnel, and quality Maintenance and material inventory management.
Process control level	Process control, HMI, and SCADA systems	Superordinate process control and monitoring (for continuous, batch, or discrete processes) Operation, monitoring, measurement archiving Recipe administration and implementation.
Automation level	PLCs, motion controllers	Control, Monitoring and diagnostics for the equipment and machines to be automated.
Field level	I /O -modules, field devices, field bus	Collection and processing of information from the technological process through sensors Active process intervention through actuators.

Initially, approaches for the decentralization of systems within factory automation occurred primarily at the hardware level based on PLC and field bus systems with process oriented automation functions (e.g., systems for decentralized periphery such as SIMATIC ET 200, or for DCS, such as PCS7). An essential objective here was the reduction of cabling and thus of the hardware cost. The consistent further development of these concepts led to intelligent field devices and the distribution of process oriented process control functions. Such decentralization, however, applied mainly to Level 1 Systems with process-oriented functions. In the meantime, numerous research projects develop decentralized concepts that include also the higher levels of the classic automation pyramid, such as production control systems and material flow computers reaching level 2 & 3 systems. The basic idea of these approaches is to modularize "central" process control tasks and to decentralize them to basic control unit types:

- Independent functional automation units representing and controlling the resources of a factory.
- Independent production process units representing and controlling the processed objects within the factory (e.g., a manufactured workpieces, products or transported goods).

Decentralized process control is achieved by situation-based interaction of controlling units of resources, as well as those of processed objects, resulting in a dynamic online setting for the process. For implementation of such approaches, mainly agent-systems are applied.

Use Cases of Decentralized Automation Systems in Factory Automation

There are various reasons for research and development of decentralized automation technologies. However, decentralized automation systems do not have an advantage per se, but must rather be regarded as means to achieve certain benefits and objectives. Examples for the benefits are:

- Increase of the flexibility and improvement of the adaptability of the automation system.
- Higher efficiency (throughput, use of resources, degree of automation).
- Reduction of communication effort.
- Increased robustness and reliability by local troubleshooting.
- Reduction of complexity for integration of resources/machines.
- Faster setup and reconfiguration.

The following two applications of decentralized automation were developed in the framework of the research projects PABADIS'PROMISE and Internet of Things. They are representative for approaches in decentralized factory automation and can be found in a similar manner in other approaches.

Flexible and Decentralized Production Systems

Accelerated innovation, shorter product life cycles, and more variants combined with small batch sizes represent new tasks for production automation. In the future, production plant suppliers must provide the following functionalities in addition to the actual automation:

- Production flexibility through individual production planning, and optimization.
- Vertical integration of information and functions at the control, MES, and ERP levels.
- Fast reconfiguration and scalable capacity through dynamic resource integration ("Plug'n'Produce").

In the European research project PABADIS'PROMISE, an architectural basis for flexible and vertically seamless production automation was developed. Key features of the approach are:

- Product driven definition of control processes instead of machine centered definition.
- Merging of the previously hierarchically separated tasks of control and MES systems.
- Decentralization of the automation functions into independent, cooperating units.

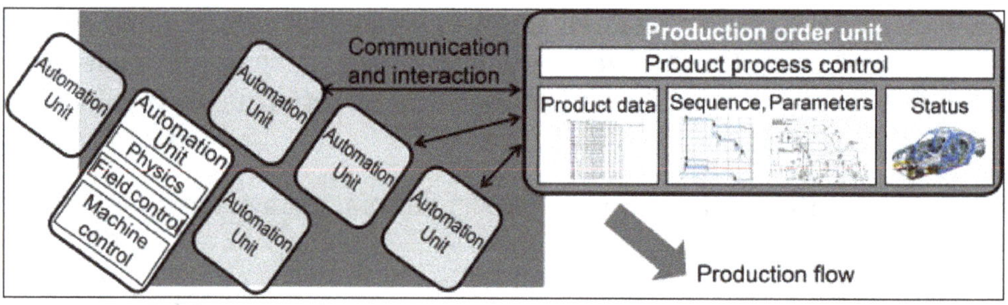

Decentralized automotive production line acc. to the PABADIS'PROMISE approach.

In this approach, the product to be manufactured represents the set point, while the control system is based on an explicit description of product data and production processes (machine independent working processes and parameters). Machine interfaces are described accordingly in a behavior-oriented manner. The integrated description model serving as basis represents an expansion of the IEC/ISO 62264 Standard. Based on these descriptions, the product-specific, on-the-fly configuration of the specific production processes of the production system and the dynamic integration of additional modules at runtime are implemented by means of software agents and RFID technology. Machines and production orders are represented by a logically and physically decentralized agent network, which coordinates the production process dynamically according to machine functions, degree of utilization, and error status, as well as allows the seamless online access from the ERP level to the field level. Figure depicts an example of decentralized control of an automotive production line where the approach was prototyped.

Flexible, Decentralized Intralogistics Systems

The term intralogistics refers to the logistical flow of goods and data “inside the four walls” of a plant. A typical example of an intralogistics system is an automatic high bay warehouse for storage and retrieval of goods in combination with an automated transport system that moves these goods within the plant. Currently, such systems are designed to fulfill a specialized and more or less static workflow. In the future, there will be a stronger demand to support or quickly adapt to different workflows to increase flexibility and adaptability both in the warehouse and in the transport area.

A practical application example is the increasing third-party business in the intralogistics area in which logistics companies perform storage and shipping services for one or more customers. Because of the mostly short contract relationship, the necessity results to adapt such a facility quickly and in an optimized manner to new customers and their requirements. Considering contemporary facility designs, this would result in frequent reconstructions. Due to high efforts, such a reconstruction is often only done in a limited manner, for which reason a part of the benefit potential of the facility will not be achieved. By means of a cost effective provision of a higher flexibility through a convertible facility, third-party logistics suppliers may achieve a significant cost advantage.

Requirements on increased flexibility and adaptability of automation solutions are for example providing additional storage capacity, supporting the storage and transport of new types of boxes, introducing new storage and picking strategies, including new sources and destinations in the transportation system. These adaptions must be supported effectively by a more efficient engineering of the related disciplines, such as mechanical, field control, and process control engineering. There is the notion that higher flexibility can be achieved through an efficient modularization of the intralogistics system, along with suitable engineering processes. The “Internet of Things” research project addresses these demands by developing a new modularization concept.

Up to now, modularization is aiming mostly at the creation of modules within each of the abovementioned disciplines, i.e., modularization is done horizontally as shown in Figure below where every discipline creates its “optimum” modules and standards that help to improve the level of integration between those discipline-specific modules. The problem here is that the integration between the systems of the different disciplines – and, therefore, the integration of the entire facility – is not addressed through such horizontal approach and, therefore, must be implemented specifically for each project. This fact leads, according to experience, to substantial additional costs, particularly during the commissioning phase.

The "Internet of Things" project introduces the concept of mechatronic modularization, stating that modularization is oriented towards the physical structure of intralogistics systems, and it is designed vertically with each module containing parts of the mechanical, field control, and process control engineering disciplines. This modularization concept is applied in a consistent and uniform manner to all engineering tasks that have to be performed during the lifecycle of an intralogistics system, in the design phase as well as in the commissioning and the reconfiguration phases.

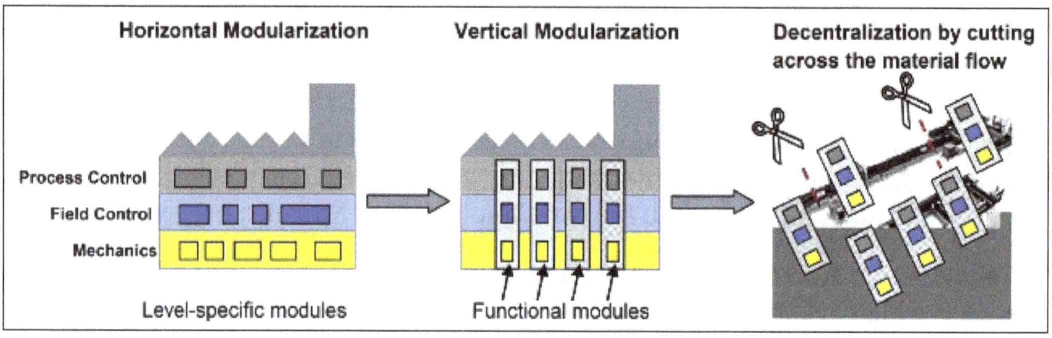

Modularization and decentralization.

An example of such a mechatronic module within the transport system could be a divert module that sends a box either straight on or causes it to turn left or right. The mechanical part of the module is the divert itself. The field control part comprises, among others, the control logic required for switching the divert to straight or left/right along with the sensors and actuators. The process control level contains routing capabilities that analyze the transport destination of the box and decide if it should move straight on or not.

Divert in baggage handling system.

In contemporary intralogistics systems, the functionality of the process control level is implemented in a Material Flow Control (MFC) System. Once this functionality is also integrated into the concept of the mechatronic modularization, it means the dissection of the previously centralized MFC System into autonomous modules. This procedure is necessary since the modules must be easily combinable with each other and, therefore, must not have interdependences with each other. Existing, possibly complex interdependences should remain encapsulated within the modules and not appear outside of the modules (cohesion). The total functionality is dissected into independent partial functionalities, which are integrated again at runtime.

By doing this, the automation pyramid in the classical sense is largely dissolved. Decisions previously taken centrally are now taken at runtime by interaction between the individual modules. The concept of the Internet of Things research project is that modules provide mechanisms for two-way identification, for coordination among each other, and with the environs, for determination of decisions, or additionally for the dynamic adaptation to changed conditions. An example is a box to be conveyed in the intralogistics system that communicates its identification at each divert and identifies its target location by RFID to the divert module. The divert module now communicates with other material handling modules in the intralogistics system, such as other diverts, conveyors, and merges, in order to determine which path the box must take to reach the target location.

The use of pre-integrated modules, the encapsulation of the functionalities, as well as the ability of the modules for identification and coordination among each other, lead to the fact, that modules can, referring to automation engineering, be integrated with each other to an entire facility on a "Plug&Play" basis. This allows supporting the mentioned requirements for flexibility regarding construction and reconfiguration of facilities.

Essential Aspects in the Life Cycle of Factory Automation Systems

For a comprehensive analysis of a decentralized automation approach, we need to map it into the context where it is applied – the industrial life cycle.

Factory automation systems – like all other industrial facilities – are complex mechatronic systems according to, consisting of the synergetic integration of different technical components and partial systems, which together fulfill a specific function. By doing so, the number of identical systems is substantially reduced for larger, more complex solutions; while devices and their components are mostly sold as ready products, factories are generally a one of a kind installation constructed under contract based on specific customer requirements. For the installation of a factory, specialists of many technical disciplines participate, such as mechanical engineering, process engineering, energy and automation engineering, informatics, etc., whose extensive knowledge and activities have to be integrated in the project.

The term Engineering is understood to represent the entire technical working process within a plant project, starting with the concept, and including Detailed Design, Realization, Installation, and Commissioning up to the transfer of the factory figure). This process contains, particularly, the selection, design, and integration of the components or partial systems of the subsections to an integrated mechatronic system. The objectives of the Engineering are: (1) to secure an error-free interaction of the integrated components relating to the overall function, and (2) the implementation in a predictable and efficient project development process.

The factory and solution business encompasses the business and technical cooperation of many participants through an often fragmented value-added chain in connection with a high share of purchased components and services that must all be integrated in the context of a project. Table shows schematically the most important stakeholders in the industrial life cycle, their particular deliverables/assets, and their different business goals. The effects of an innovative automation concept must be considered separately for each player.

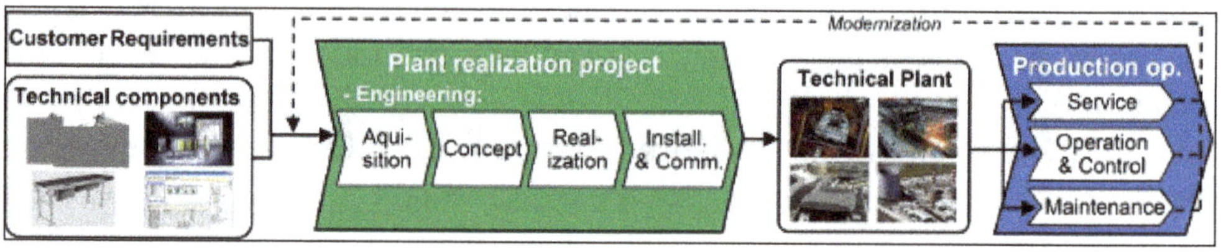

Engineering of plants: Integration of individual components to an entire plant.

It is important to notice that decentralized automation does not only affect the technology, but also changes the respective planning processes and models. The value-adding parameters in the engineering phases cannot be found in the technology itself, but in their application in plant design and implementation. The most important challenges in plant engineering are:

- More efficient development and faster time-to-operate by efficient work share, interdisciplinary cooperation, and continuity in the life cycle.
- Reduction of project risk and safeguarding of the project through systematic engineering and continuous validation throughout the project, as well as through cooperation and exchange of information in the supply chain.
- Effect of quantity and increase of quality through repetitive use of modular solutions.

Consequently, we identified the need to develop appropriate engineering methods and processes for decentralized factory automation systems and to show how they address the challenges listed above.

Table: Stakeholders in the Industrial Life Cycle.

Stake-holder	Component Supplier	Equipment/M a chine Supplier	System Integrator	Plant Operator/ Owner
Description	Provides automation technology, devices, & components	Builds machines/ plant equipment by using components	Builds plants, integrates equipment and components	Designs products, defines production requirements & workflows
Deliverables/Assets	Sensors, drives, PLC, panels, control and MES systems	Robots, CNC or assembly machines, conveyor systems	Technical plant & automation solution	Product model, production process, rough material flow
Example Depiction of Deliverables/Assets				

Business Goals	Value added by flexible, integrable components and systems	Value added by integration of components to marketable machines	Value added by specific integration of systems, equipment & components	Flexible, optimized production process, availability and capacity utilization
Scope	Make-to-Stock, large numbers Wide application scope	Make-to-Stock/ Adapt-to-Order Domain scope	Make-to-Order Project-oriented	Production scope (ERP to production execution level)

Engineering of Decentralized Factory Automation Systems

Engineering of a Production Line

When working on decentralized solutions for automated factory systems, the question was how to plan, design and implement such decentralized automation solutions. For the solution of this problem, an engineering model was developed that describes the possible processes for the engineering of decentralized production automation systems under consideration of mechatronic modules. In order to provide a reference, existing engineering projects were analyzed first and a generalized engineering process for factory automation was documented with basic tasks described in table.

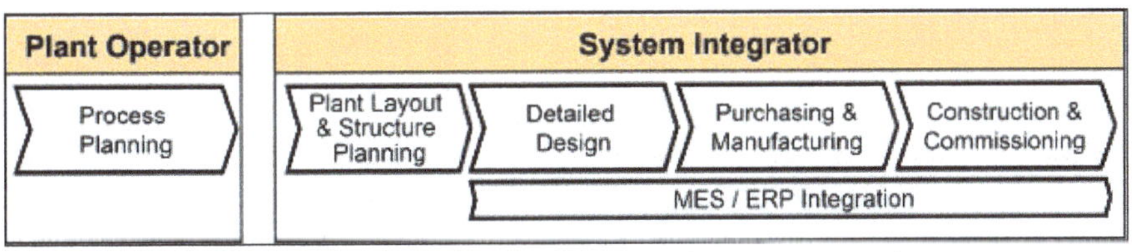

Generalized engineering process for factory automation.

Table: Phases and tasks of the basic process for manufacturing system engineering.

Phase	Basic Engineering Tasks
Process Planning	Define manufacturing steps and their physical and logical order.
Plant Layout & Structure Planning	Derive plant layout from the process description Specify machine types by means of required manufacturing capabilities Specify material transport.
Detailed Design	Specify technical details of plant equipment, instrumentation, and automation (parallel proceeding of all involved technical disciplines) Automation engineering: extend layout by control devices, detail production steps, and implement control.
Purchasing & Manufacturing	Purchase/manufacture devices, equipment, and machines Construct and implement toward functional units (all involved technical disciplines).
Construction & Commissioning	Integrate functional units toward complete plant Verify and optimize machine and process capability.
MES /ERP Integration	MES: Specify the interfaces between control system and MES system; configure MES system based on the layout and the manufacturing process ERP: specify the data for exchange; implement the necessary interfaces and communication.

Engineering of a Decentralized Production Line

For the case of the development of a facility with decentralized automation, the previously connected automation functionality, in the course of engineering, must be distributed to the individual technical resources (machines, etc.). For an efficient design of the engineering process, these modules must integrate all system specific parts of a facility resource, as well as must be used consistently. In order to avoid the effort for the specific dissection of the facility into individual mechatronic modules for each individual project, these modules must be specified in advance and provided as comprehensively as possible.

Table: Engineering of a Decentralized Production Line.

Plant Operator	System Integrator		
Process Planning	Plant Layout and Structure Planning	Detailed Design	Construction & Commissioning
Use of pre-defined generic production activities Define hierarchies and dependencies	Use of a pool of predefined mechatronic modules Match available modules with required production activities	Control engineering: Configure module functionality Program additional control functionality	Control integration via Plug-and-Play Virtual commissioning

Process Planning

The production process for the product to be produced is described by a specification of the production steps by the operator of the production. In addition, the production process has to be structured hierarchically in production steps, and the predecessor-successor relation, as well as parallel processing between production steps, must be described. Likewise, the machine functionalities for the performance of a production step have to be characterized. This concerns both the control functionality and the kinematic behavior of machines. All mentioned additional steps serve for the preparation of the efficient implementation of the subsequent steps.

Plant Layout and Structure Planning

Based on these information's and the available mechatronic modules, the system integrator derives the facility parts to be installed and their technical interrelationship. Mechatronic modules that offer the desired functionalities are attributed to each production step. For this purpose, the mechatronic modules will be selected from a library that may also contain complete process step realizing production cells. For selection and matching, the predefined production functions of the mechatronic modules are used. If the production process was specified comprehensively as described in Process Planning, the Plant Layout and Structure Planning may be performed in a supported or automated manner.

In addition to the pre-specified, mechanical and electronic data, the mechatronic module also contains predefined control applications for the realisation of the production functions. In addition, the specific data of the utilized mechatronic modules for the geometrical positioning in the facility, their kinematic behavior, and their geometries in the 3D factory layout has to be substantiated. Furthermore, the interdependence is considered for the dimensioning of mechanics and electrics, e.g., for the layout of the power of motors.

Detailed Design

In the next step, the configuration of the predefined control functionalities of the utilized mechatronic modules is done in accordance with the specific process parameters, and, if necessary, the programming of auxiliary functions is performed.

In addition, the connections between mechatronic modules and the corresponding parameterizations of these connections have to be specified:

- Connection of electrical, pneumatic, and hydraulic systems.
- Connection of the communication devices to communication networks.
- Connection of material flow relations.

The connections are generated automatically or manually, depending on to what extent the connections have been specified in advance.

PLC and HMI programs are usually generated automatically from the control applications of mechatronic modules. In this planning phase, the design or the adaptation of the facility mechanics (CAD) and the electrical design (CAE) are also done unless large extends have been already predefined in the mechatronic modules.

MES/ERP Integration

In this planning phase, initially, the interfaces between the predefined process control and MES functions of the mechatronic modules on the one side and the facility MES level on the other side are designed. Engineering data that have to be considered in the design of the MES modules and the facility MES functions are:

- The production process with its specified production steps.
- The facility layout.
- The technical relations of the mechatronic modules in the facility.

Once this information is available in a comprehensive manner, the essential tasks can be partially automated. Additionally, control applications have to be adapted to specific MES functionalities – for instance, the scheduling of production – if support still exists. Once the MES functionalities are designed, the parameterization of the interfaces to the ERP system and the synchronization of relevant data between the MES and the ERP system are performed.

Construction and Commissioning

As the real used facility resources correspond to pre-integrated mechatronic modules, they can be provided in a manner in which they can be integrated into the automation system via plug and play. After corresponding adaptation, these facility parts can be used immediately.

Predevelopment of Mechatronic Modules

The essential precondition for the feasibility of the engineering of production facilities according to the described procedure is that, in the run-up, mechatronic modules are prepared in order to be available as a template in a library for the facility planner. This task requires an additional project independent phase and corresponds to a predevelopment.

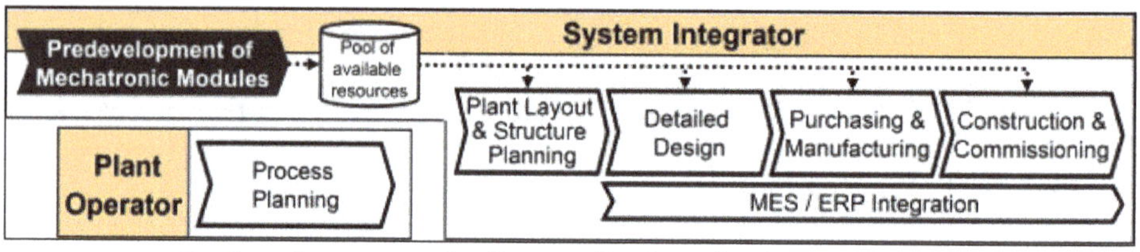

Engineering process for decentralized production automation systems based on mechatronic modules.

During the design of these facility and project independent mechatronic modules, the following information must be specified and provided in an integrated manner:

- Description of the mechanical parts, e.g., in CAD layout.
- Description of the behaviors of the facility resource, such as kinematic behavior.
- Description of the machine functionalities that the resource offers for the design of production steps (function oriented view).

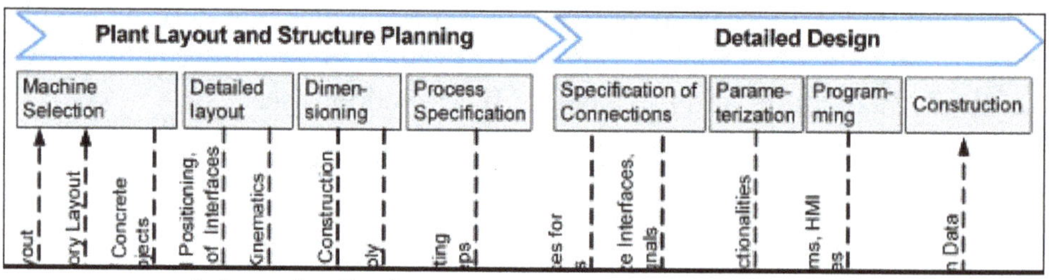

Information flow between engineering tasks and modules.

- Description of the automatisation portions: control and communication interfaces, required connections for hydraulics, pneumatics, and electrics, as well as models or implementations of baseline control, HMI, and communication building blocks.
- Description of combination possibilities or limitations with other resources in order to combine individual functionalities into one overall functionality.

The information flow for two of the abovementioned phases, which converges in an orderly fashion using mechatronic modules.

Engineering of a Material Flow System

As for the description of the production line scenario, for the engineering of an intralogistics material flow system a general engineering process is described before a model is presented that offers a possibility for the engineering of the decentralized material flow system. The phases refer to the partitioning of the processes as presented in figure below:

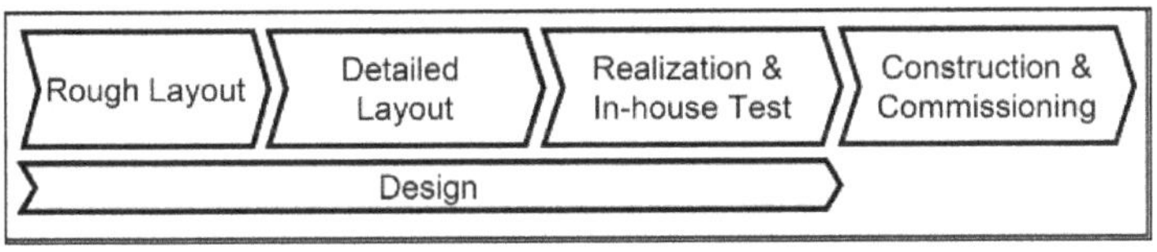

Generalized engineering process for intralogistics systems.

Table: Phases and tasks of basic process for engineering of material flow system.

Phase	Basic Engineering Tasks
Plant Topology and Rough Layout	Deduction of the rough layout based on the required material flow (requirement, particularly in view of transport functionalities, storage, handling, throughput of the factory), rough simulation for the detection of bottlenecks.
Detailed Layout and Specifications	Refinement and detailed specification of the requirements, establishment of detailed layout, adaptations, optimization of parts lists, concept for electrics, determination of the IT concept, strategies for control algorithms, definition of the interfaces between disciplines, refined simulation based on control algorithms.
Realization and In-House Test	Procurement, partial installation, PLC, and IT side configuration and programming related individual tests/ interface tests automation/IT.
Construction/Commissioning	Installation of the systems, commissioning of control system, test of IT interfaces with other processes.

Engineering of a Decentralized Material Flow System

The request for flexible integration and reconfiguration of an intralogistics system could be realized by the creation and continuous use of mechatronic modules. For this purpose, previously centralized automation functionalities are distributed to individual modules, which lead to the decentralization of the intralogistics system.

This approach has implications on the architecture of the automation system and, therefore, on the handling of the modules. This is reflected by the introduction of an engineering process for decentralized systems that supports the engineering in the design, commissioning, and reconfiguration phases. This will be presented hereafter.

Plant Topology and Rough Layout

As common, a rough layout is created, based on the customer requirements and the specifications of the factory planners, that describes the general topology of the system and all source/target relations. From this layout, the system integrator identifies the factory parts to be installed and their technical interrelations as depicted in figure. These factory parts may consist of one or more mechatronic modules. One example for a factory part consisting of several modules is a buffer storage unit for intermediate storage of boxes, consisting of the material handling modules, a rack, and an automated storage and retrieval system (AS/RS).

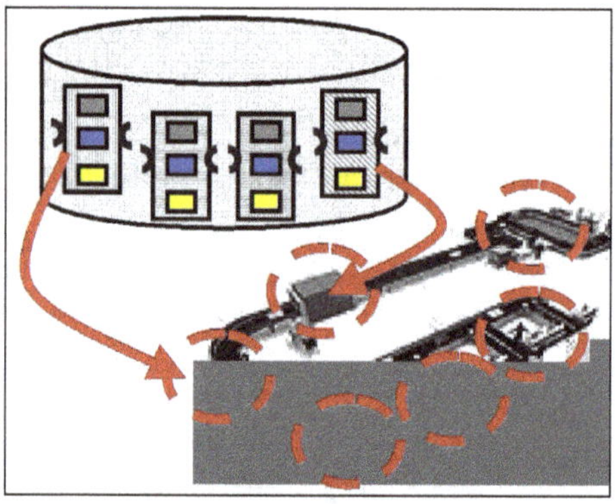

Identification of appropriate modules.

In this phase, not only the factory components but also the mechatronic modules are still generic. For example, it is determined where the buffer storage unit is located in the topology, how many spaces it must have, which throughput it performs, and which functions are generally required. But it is not yet defined which particular conveyor types and which AS/RS will be used. At this time, a rough simulation of the process is performed for the verification of the throughput.

Detailed Layout and Specification

Now, a successive refinement of the developed requirements and definitions is done in a top-down approach. Initially, the requirements for the mechatronic modules are further detailed. Then, it is determined which mechatronic modules will be used in reality, e.g., the conveyor of Company A, that can move a certain maximum load, and a rack feeder of Company B. At this time, the identification of components to be developed specifically for the project is also made, such as, for example, a special gripping device for the rack feeder for the nonstandard boxes of the customer, including the corresponding automation system.

In this phase, the control and process concepts are specified. In opposite to central automation, in which the process logic can be explicitly specified, the procedures may only be defined through the rules of interrelations. As an example, the control of a transport process by driverless transport systems (Automated Guided Vehicles, AGV) can be considered. This control can be implemented by a framework of auctions. A bid for the box transport can be requested by the box while the AGVs will provide it based on availability of the corresponding AGV, its actual location, its transport capacity, and other parameters. The box transport than shall be assigned to the AGV with the most "beneficial" bid. That means that the engineering process is performed at a more abstract level; the possibilities of interactions of the modules must be very familiar to the developer. Then, a refined simulation of the processes is performed. This simulation exploits functionalities of the mechatronic modules used later on in the actual operation in suitable granularity introducing them into the simulation by Plug&Play. This simulation allows the rough testing of the effects of different control concepts (i.e., different system configurations). This also allows identifying components and rules to be developed specifically for the project.

The interfaces between the modules are defined using standardized interface descriptions. The interfaces of the mechatronic material flow system to superimposed processes (e.g., ERP system or Warehouse Management System) and neighboring processes (e.g., the production control system) are defined. In addition, the connections of the mechatronic modules to the central infrastructure components are defined, such as, e.g., electrical power supply, compressed air supply, and connection to the IT network.

Realization and In-house Test

The individual mechatronic modules are configured and parameterized for their particular task in accordance with the requirements of the established specifications. For this purpose, the specialists of the different technical disciplines continue to work on their specific tasks. However, its work is based on consistent mechatronic modules that provide the corresponding views. The necessary project-specific designs are performed. Based on the control concept and the defined possibilities for communication and coordination between the modules, the concrete control mechanisms are configured and project-specifically extended if necessary. The functionality per se is generated at runtime by using defined rules.

An in-house test of the system is performed during which a pretest of the facility is done to the possible extent. The vertically modularized structure and the "Plug&Play" functionality support virtual commissioning by subjecting readily configured or installed facility parts to the test and recreate others that are not completed by simulation or emulation.

Construction/Commissioning

During the commissioning phase, the "Plug&Play" functionality, i.e., the use of pre-integrated modules with the capability to identify each other and coordinate among each other, is applied. It allows a bottom-up installation of the factory according to the factory layout. The physical modules possess mechanisms in order to recognize at which position of the factory topology they are located and perform their configuration and parameterization based on their specifications. They connect themselves with their neighboring modules in terms of automation and communicate with each other.

The use of pre-integrated mechatronic modules allows a functionally oriented installation of the factory and, therefore, an early testing of interconnected partial areas and, specifically, an early start of the commissioning of the automation. It is not necessary to integrate completely mechanics and electrics before commissioning the automation, but it is possible to combine and test small functional units such as, for instance, a material handling circuit that is able to move boxes while additional integration still occurs around it.

Simplified Integration.

Reconfiguration

Even in existing plants, reconfiguration procedures occur. They generally require extensive engineering processes. If, for instance, additional diverts shall be included in the automatic transport system in order to be able to move to different locations, an entire engineering process needs to be performed for such a change. Then, the mechanics of the transport system must be separated at the plant locations in the context of the commissioning, the diverts must be mechanically installed, and subsequently, the electrics and the automation system must be installed and newly commissioned.

Simplified Adaption to Changes.

In the case of the decentralized mechatronic system, these activities are supported by the "Plug&Play" functionality of the modules. Modules can be removed from the overall system without causing far-reaching changes to the automation system (dynamic adaptation to changed topology) as shown in figure. In terms of automation, new modules, to a large extent, are autonomously integrated into an existing plant. The required extent of the engineering depends essentially, in this example, on the requirements for actuality and consistency in the plant documentation, such as, for example, the updating of the plant layout.

Predevelopment of Mechatronic Modules

In order to keep the project specific effort for the use of these modules as low as possible, they must be defined and specified as far as possible in advance. This means the creation of mechatronic module building blocks that are defined independent of specific projects. These modules provide all involved disciplines with their particular views using a consistent model. The building block approach supports the top down proceeding during the design phase by the provisioning of module descriptions based on which the initially still rough plant layout can be refined successively. Therefore, standardized means of descriptions must be applied. They contain required module information related to the different disciplines or to the proposed (transport) services, for instance.

There are functionalities in an intralogistics system that require comprehensive information, such as, the routing of a box to its target destination or the allocation of plant resources to tasks based on the current load of the system (e.g., selection of automated storage/retrieval system, AS/RS). They were previously performed through central systems. Such functionalities have to be replaced for decentralized plants by other mechanisms. This includes concepts for the substitution and distribution of such control logics, capabilities for communication, as well as for coordination of the modules, such as auctions. For instance, the decision finding for allocating a storage order to an AS/RS, is done rule-based in the decentralized case. Formerly, a central system maintained information about the load of the AS/RS and made decisions for a suitable device. Now, whether an AS/RS may offer its services or not has to be determined by rules. In addition, it must be determined according to which rules the devices coordinate among each other about the allocation of an order and, lastly, select one of them for the execution of the storage order.

Methodical Evaluation of Benefits and Risks

Changes, such as the introduction of mechatronic modules, a mechatronic approach in engineering, distributed control logic, and function-oriented commissioning, are only supportable if their impact, in terms of benefits and risks, on the entire industrial life cycle, has first been worked out in detail.

For the planning, realization, and commissioning phases, there is no well-known procedure through which the added value of an automation concept can be determined. Regarding the introduction of new concepts, it is of high importance to evaluate benefits and risks before the actual practical deployment. Although an effort-and-cost based evaluation would be desirable, the lack of completed projects and reliable figures prevents the use of such an evaluation. In order to close this gap, an evaluation strategy was developed the PABADIS'PROMISE and "Internet of Things" research projects. The basic idea of the method is the creation of reference models for engineering processes and activities along the life cycle of an automation solution. These models refer to the corresponding automation concepts, for instance, the classically developed system and the decentralized automation system developed mechatronicaly, to compare both.

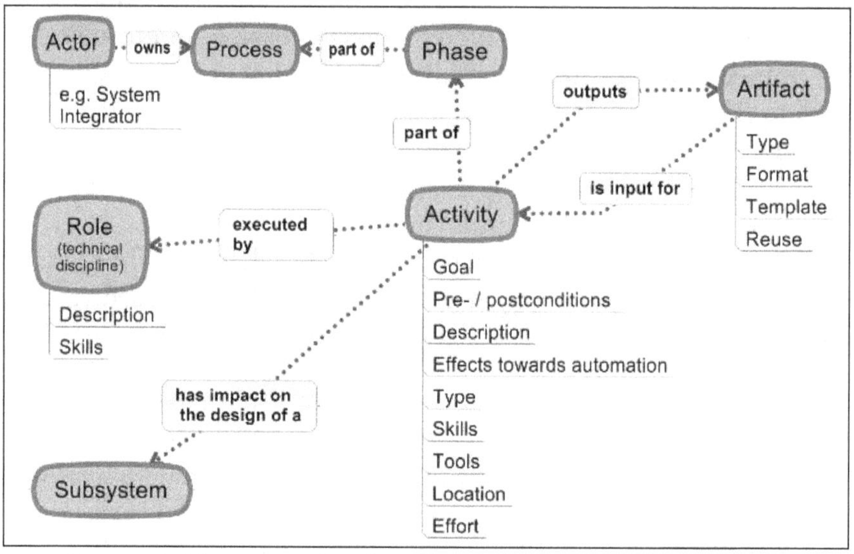

Meta-model for structural and attributive features of engineering task.

The creation of such a model is done by dissecting the life cycle of the plant in its phases by the dissection of the processes (e.g., layout planning) within the individual phases into individual activities and by the identification of their relation with the automation system.

For the creation of a neutral reference basis for the later comparison of different automation concepts, a meta-model was developed which defines all required features of the corresponding reference models for engineering and commissioning in a uniform manner. As an auxiliary tool for building the reference models, corresponding templates were created to record activities/processes.

The evaluation is done by a comparative assessment. As one of the basics depicted in table, generally applicable evaluation criteria were defined, derived from central levers in factory engineering, such as modularization, re-use, integration, and seamlessness, as well as under the viewpoint of applicability. They serve as guiding issues for the identification of critical factors in engineering and address specifically the challenges concerning the phases. In addition, the standard ISO/IEC 9126, which regards the assurance of software quality, was evaluated, and the quality characteristics included therein were transferred, adapted, and amended to meet the requirements of plant business. The criteria catalogue thusly generated serves as the basis for the evaluation; an excerpt of questions can be found in table.

Table: Excerpt of criteria catalogue.

Topic	Criteria	Question	Details
Suitability	Customer requirements	Which customer requirements are relevant for the system integrator in order to design and realize the plant?	Typical customer requirements today aims of system Control architecture specific requirements.
	Appropriateness	Are the requirements considered in a suitable way?	e.g., highly redundant (and costly) automation system in case of non-time-critical processes is not appropriate.
Interoperability	Content- related dependencies between activities	Which dependencies during the engineering process do exist between automation engineering and other disciplines?	Required information from mechanical engineering, electrical engineering and mechanical plant layout processes to be controlled/monitored, list of process signals sensors/variables to measure the process actuators/variables to influence the process.
	View integration	How is cooperation of technical disciplines supported?	e.g., integrated information models such as multi-discipline information objects

For the automation concepts to be compared, according to the evaluation proceeding a reference model is initially established for the processes and activities along the life cycle. The information serves as a basis for the creation of these models for the presented automation solutions. Subsequently, the different models are analyzed based on the evaluation criteria and compared to each other related to benefits and disadvantages.

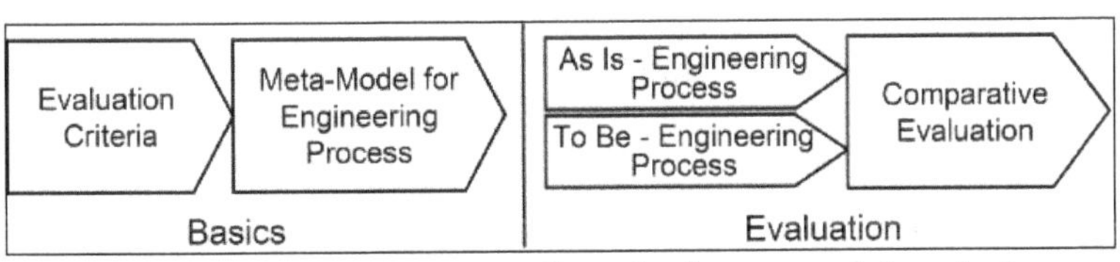

Basics and evaluation proceeding of reference model method.

Implications for the Engineering Processes

The detailed gathering and analysis of the individual processes and activities for both decentralized production systems and, also, decentralized material flow systems showed very similar results as compared to previous engineering processes. To begin with, the following findings are related to the activities to be performed and the resulting (expected) efforts.

Predevelopment

For the creation of mechatronic blocks, a correspondingly higher effort has to be undertaken for predevelopment. This includes:

- Adequate definition of standard modules with as few as possible technical and functional dependencies by systematic dissection of a plant.
- Coordination across the involved disciplines for the generation of mechatronic modules.
- Standardized interface descriptions (functional, technical, information technical).
- Standardized description of process functions or partial processes provided by a module.
- Description of sequences of processes and the invocation of process functions.
- Provision of configuration possibilities and interfaces for project specific adaptations.
- Provision of module-specific automation control functionality and its implementation.
- Provision of application guidelines for modules (e.g., application areas, limitations, dependency on configuration possibilities, and variants).

The creation process for mechatronic module building blocks has to be done in cooperation and coordination with all participating stakeholders (system integrator, equipment/machine supplier, and component supplier).

References

- An-overview-on-industrial-automation: elprocus.com, Retrieved 16 June, 2020
- Gilchrist, Alasdair (2016). "Industry 4.0 - the industrial internet of things". Apress Media. doi:10.1007/978-1-4842-2047-4. ISBN 978-1-4842-2046-7
- Benefits-industrial-automation: thesunflowerlab.com, Retrieved 05 August, 2020
- "How infected IoT devices are used for massive DDoS attacks - Fedscoop". Fedscoop. 26 September 2016. Retrieved 11 May 2017
- Automatic-test-equipment-ate-2148: techopedia.com, Retrieved 14 March, 2020
- "Cyberthreats, Vulnerabilities and Attacks on SCADA Networks" (PDF). Rosa Tang, berkeley.edu. Archived from the original (PDF) on 13 August 2012. Retrieved 1 August 2012
- Advantages-of-industrial-automation-with-robots: robots.com, Retrieved 16 February, 2020
- Engineering-processes-for-decentralized-factory-automation-systems, factory-automation: intechopen.com, Retrieved 24 April, 2020

3

Control Systems in Automation

Control systems deal with the operation of control systems and instrumentation used in different processes related to automation. It includes discrete control, PID controller, sequential control, motion control, distributed control system, etc. The topics elaborated in this chapter will help in gaining a better perspective of these control systems in automation.

Discrete Control

Discrete control is one of the major subsystems in the industrial automation and needs of the manufacturing sectors are the primary driver for the growth and maturity. It is also called as two- position control system that has only two states fully on and fully off. It is the oldest type of control strategy that gives simple on-off mechanism for the set limits. When the process variable (PV) or measurement is below the set limit, then the controller is switched ON and the manipulated variable (MV) changes to maximum value. Similarly, if the PV is above the set limit, then the controller gets turned OFF and the MV changes to minimum value.

But, practically ideal on-off controller is not recommended since it is subject to process disturbances and other electrical interferences. To prevent this, on–off hysteresis or differential is added to the control function.

Open-loop Control

In this control structure, control action is not a function of process variable or any load changes. This is also not a self-correcting one. This control scheme independently calculates control output needed to the actuator to produce the desired response. But this system never knows the status of actuator output or process condition and also controller doesn't consider any disturbance inputs in the process. That's why this is limited for most of the applications.

Feed-forward Control

This control scheme is implemented to reduce the effect of input disturbance on process. A sensor in this system detects the disturbances or any load changes in the process, and then controller calculates the necessary signal to the actuator. This allows the controller to take corrective action before the disturbance effect the process. But this control scheme is more expensive and complicated and requires a better understanding of the process to implement control logic.

Closed-loop Control

Closed-loop control scheme is widely implemented in many of the industrial control applications. It uses the output measurement as feedback signal; and, it is compared with desired or set-point value. The difference between these actual and desired output is called error. Thus, the controller gives the control input to the actuator till the error gets minimized so that the output is maintained at the desired value. Most of the closed loop control systems are implemented with PID controller for accurate and reliable control.

Sometimes, the control element has only two positions either it is fully closed or fully open. This control element does not operate at any intermediate position, i.e. partly open or partly closed position. The control system made for controlling such elements is known as on off control theory. In this control system, when process variable changes and crosses certain preset level, the output valve of the system is suddenly fully opened and gives 100 % output.

Generally in on off control system, the output causes change in process variable. Hence due to effect of output, the process variable again starts changing but in reverse direction. During this change, when process variable crosses certain predetermined level, the output valve of the system is immediately closed and output is suddenly reduced to 0%.

As there is no output, the process variable again starts changing in its normal direction. When it crosses the preset level, the output valve of the system is again fully open to give 100% output. This cycle of closing and opening of output valve continues till the said on-off control system is in operation.

A very common example of on-off control theory is fan controlling scheme of transformer cooling system. When transformer runs with such a load, the temperature of the electrical power transformer rises beyond the preset value at which the cooling fans start rotating with their full capacity. As the cooling fans run, the forced air (output of the cooling system) decreases the temperature of the transformer. When the temperature (process variable) comes down below a preset value, the control switch of fans trip and fans stop supplying forced air to the transformer.

After that, as there is no cooling effect of fans, the temperature of the transformer again starts rising due to load. Again when during rising, the temperature crosses the preset value, the fans again start rotating to cool down the transformer. Theoretically, we assume that there is no lag in the control equipment. That means, there is no time day for on and off operation of control equipment. With this assumption if we draw series of operations of an ideal on off control system, we will get the graph given below.

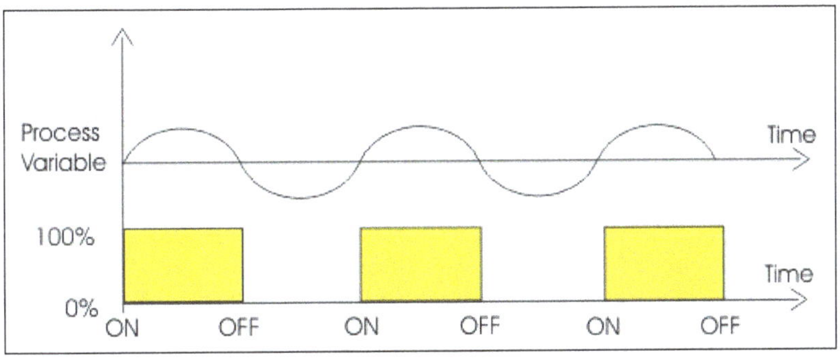

But in practical on off control, there is always a non zero time delay for closing and opening action of controller elements. This time delay is known as dead time. Because of this time delay the actual response curve differs from the above shown ideal response curve. Let us try to draw actual response curve of an on off control system.

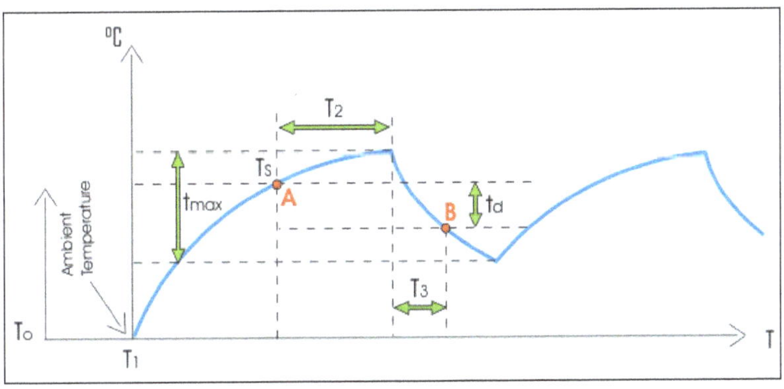

Say at time T_0 the temperature of the transformer starts rising. The measuring instrument of the temperature does not response instantly, as it requires some time delay for heating up and expansion of mercury in temperature sensor bulb say from instant T_1 the pointer of the temperature indicator starts rising. This rising is exponential in nature. Let us at point A, the controller system starts actuating for switching on cooling fans and finally after period of T_2 the fans starts delivering force air with its full capacity. Then the temperature of the transformer starts decreasing in exponential manner.

At point B, the controller system starts actuating for switching off the cooling fans and finally after a period of T_3 the fans stop delivering force air. Then the temperature of the transformer again starts rising in same exponential manner.

PID Controller

A proportional–integral–derivative controller (PID controller or three-term controller) is a control loop mechanism employing feedback that is widely used in industrial control systems and a variety of other applications requiring continuously modulated control. A PID controller continuously calculates an error value $e(t)$ as the difference between a desired setpoint (SP) and a measured process variable (PV) and applies a correction based on proportional, integral, and derivative terms (denoted P, I, and D respectively), hence the name.

In practical terms it automatically applies accurate and responsive correction to a control function. An everyday example is the cruise control on a car, where ascending a hill would lower speed if only constant engine power were applied. The controller's PID algorithm restores the measured speed to the desired speed with minimal delay and overshoot by increasing the power output of the engine.

The first theoretical analysis and practical application was in the field of automatic steering systems for ships, developed from the early 1920s onwards. It was then used for automatic process control in the manufacturing industry, where it was widely implemented in pneumatic, and then electronic, controllers. Today the PID concept is used universally in applications requiring accurate and optimised automatic control.

Fundamental Operation

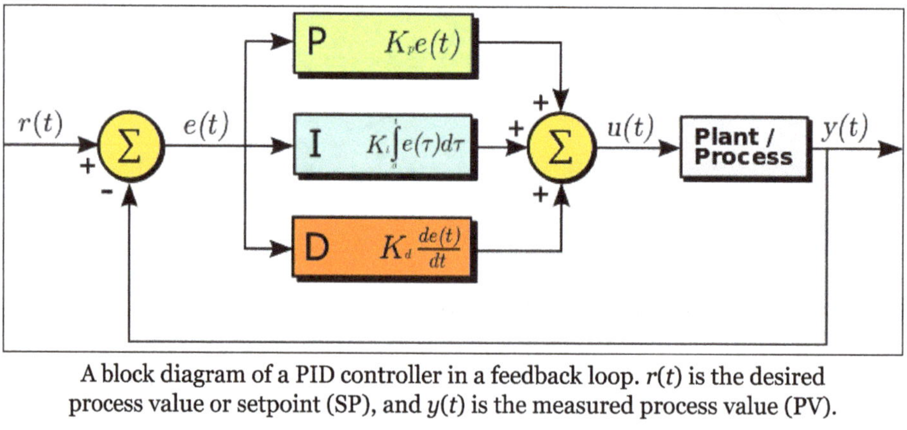

A block diagram of a PID controller in a feedback loop. $r(t)$ is the desired process value or setpoint (SP), and $y(t)$ is the measured process value (PV).

The distinguishing feature of the PID controller is the ability to use the three control terms of proportional, integral and derivative influence on the controller output to apply accurate and optimal control. The block diagram on the right shows the principles of how these terms are generated and applied. It shows a PID controller, which continuously calculates an error value $e(t)$ as the difference between a desired setpoint $SP = r(t)$ and a measured process variable $PV = y(t)$, and applies a correction based on proportional, integral, and derivative terms. The controller attempts to minimize the error over time by adjustment of a control variable $u(t)$, such as the opening of a control valve, to a new value determined by a weighted sum of the control terms.

In this model:

- Term P is proportional to the current value of the SP – PV error $e(t)$. For example, if the error is large and positive, the control output will be proportionately large and positive, taking into account the gain factor "K". Using proportional control alone will result in an error between the setpoint and the actual process value, because it requires an error to generate the proportional response. If there is no error, there is no corrective response.
- Term I accounts for past values of the SP – PV error and integrates them over time to produce the I term. For example, if there is a residual SP – PV error after the application of proportional control, the integral term seeks to eliminate the residual error by adding a control effect due to the historic cumulative value of the error. When the error is eliminated, the integral term will cease to grow. This will result in the proportional effect diminishing as the error decreases, but this is compensated for by the growing integral effect.
- Term D is a best estimate of the future trend of the SP – PV error, based on its current rate of change. It is sometimes called "anticipatory control", as it is effectively seeking to reduce the effect of the SP – PV error by exerting a control influence generated by the rate of error change. The more rapid the change, the greater the controlling or dampening effect.

Tuning

The balance of these effects is achieved by loop tuning to produce the optimal control function. The tuning constants are shown as "K" and must be derived for each control application, as they depend on the response characteristics of the complete loop external to the controller. These are dependent on the behaviour of the measuring sensor, the final control element (such as a control valve), any control signal delays and the process itself. Approximate values of constants can usually be initially entered knowing the type of application, but they are normally refined, or tuned, by "bumping" the process in practice by introducing a setpoint change and observing the system response.

Control Action

The mathematical model and practical loop above both use a "direct" control action for all the terms, which means an increasing positive error results in an increasing positive control output for the summed terms to apply correction. However, the output is called "reverse" acting if it is necessary to apply negative corrective action. For instance, if the valve in the flow loop was 100–0% valve opening for 0–100% control output – meaning that the controller action has to be reversed. Some process control schemes and final control elements require this reverse action. An example would be a valve for cooling water, where the fail-safe mode, in the case of loss of signal, would be 100% opening of the valve; therefore 0% controller output needs to cause 100% valve opening.

Mathematical Form

The overall control function $u(t) = K_p e(t) + K_i \int_0^t e(t')dt' + K_d \frac{de(t)}{dt}$,

where K_p, K_i, and K_d, all non-negative, denote the coefficients for the proportional, integral, and derivative terms respectively (sometimes denoted P, I, and D).

In the standard form of the equation, K_i and K_d are respectively replaced by K_p / T_i and $K_p T_d$; the advantage of this being that T_i and T_d have some understandable physical meaning, as they represent the integration time and the derivative time respectively.

$$u(t) = K_p \left(e(t) + \frac{1}{T_i} \int_0^t e(t')dt' + T_d \frac{de(t)}{dt} \right)$$

Selective use of Control Terms

Although a PID controller has three control terms, some applications need only one or two terms to provide appropriate control. This is achieved by setting the unused parameters to zero and is called a PI, PD, P or I controller in the absence of the other control actions. PI controllers are fairly common in applications where derivative action would be sensitive to measurement noise, but the integral term is often needed for the system to reach its target value.

Applicability

The use of the PID algorithm does not guarantee optimal control of the system or its control stability. Situations may occur where there are excessive delays: the measurement of the process value is delayed, or the control action does not apply quickly enough. In these cases lead–lag compensation is required to be effective. The response of the controller can be described in terms of its responsiveness to an error, the degree to which the system overshoots a setpoint, and the degree of any system oscillation. But the PID controller is broadly applicable, since it relies only on the response of the measured process variable, not on knowledge or a model of the underlying process.

Control Loop Example

Consider a robotic arm that can be moved and positioned by a control loop. An electric motor may lift or lower the arm, depending on forward or reverse power applied, but power cannot be a simple function of position because of the inertial mass of the arm, forces due to gravity, external forces on the arm such as a load to lift or work to be done on an external object.

- The sensed position is the process variable (PV).
- The desired position is called the setpoint (SP).
- The difference between the PV and SP is the error (e), which quantifies whether the arm is too low or too high and by how much.
- The input to the process (the electric current in the motor) is the output from the PID controller. It is called either the manipulated variable (MV) or the control variable (CV).

By measuring the position (PV), and subtracting it from the setpoint (SP), the error (e) is found, and from it the controller calculates how much electric current to supply to the motor (MV).

Proportional

The obvious method is proportional control: the motor current is set in proportion to the existing error. However, this method fails if, for instance, the arm has to lift different weights: a greater weight needs a greater force applied for a same error on the down side, but a smaller force if the error is on the upside. That's where the integral and derivative terms play their part.

Integral

An integral term increases action in relation not only to the error but also the time for which it has persisted. So, if applied force is not enough to bring the error to zero, this force will be increased as time passes. A pure "I" controller could bring the error to zero, but it would be both slow reacting at the start (because action would be small at the beginning, needing time to get significant) and brutal (the action increases as long as the error is positive, even if the error has started to approach zero).

Derivative

A derivative term does not consider the error (meaning it cannot bring it to zero: a pure D controller cannot bring the system to its setpoint), but the rate of change of error, trying to bring this rate to zero. It aims at flattening the error trajectory into a horizontal line, damping the force applied, and so reduces overshoot (error on the other side because too great applied force). Applying too much impetus when the error is small and decreasing will lead to overshoot. After overshooting, if the controller were to apply a large correction in the opposite direction and repeatedly overshoot the desired position, the output would oscillate around the setpoint in either a constant, growing, or decaying sinusoid. If the amplitude of the oscillations increases with time, the system is unstable. If they decrease, the system is stable. If the oscillations remain at a constant magnitude, the system is marginally stable.

Control Damping

In the interest of achieving a controlled arrival at the desired position (SP) in a timely and accurate way, the controlled system needs to be critically damped. A well-tuned position control system will also apply the necessary currents to the controlled motor so that the arm pushes and pulls as necessary to resist external forces trying to move it away from the required position. The setpoint itself may be generated by an external system, such as a PLC or other computer system, so that it continuously varies depending on the work that the robotic arm is expected to do. A well-tuned PID control system will enable the arm to meet these changing requirements to the best of its capabilities.

Response to Disturbances

If a controller starts from a stable state with zero error (PV = SP), then further changes by the controller will be in response to changes in other measured or unmeasured inputs to the process that affect the process, and hence the PV. Variables that affect the process other than the MV are known as disturbances. Generally, controllers are used to reject disturbances and to implement setpoint changes. A change in load on the arm constitutes a disturbance to the robot arm control process.

Applications

In theory, a controller can be used to control any process that has a measurable output (PV), a known ideal value for that output (SP), and an input to the process (MV) that will affect the relevant PV. Controllers are used in industry to regulate temperature, pressure, force, feed rate, flow rate, chemical composition (component concentrations), weight, position, speed, and practically every other variable for which a measurement exists.

PID Controller Theory

The PID control scheme is named after its three correcting terms, whose sum constitutes the manipulated variable (MV). The proportional, integral, and derivative terms are summed to calculate the output of the PID controller. Defining $u(t)$ as the controller output, the final form of the PID algorithm is,

$$u(t) = MV(t) = K_p e(t) + K_i \int_0^t e(\tau)d\tau + K_d \frac{de(t)}{dt},$$

where,

K_p is the proportional gain, a tuning parameter,

K_i is the integral gain, a tuning parameter,

K_d is the derivative gain, a tuning parameter,

$e(t) = SP - PV(t)$ is the error (SP is the setpoint, and PV(t) is the process variable),

t is the time or instantaneous time (the present),

τ is the variable of integration (takes on values from time 0 to the present t).

Equivalently, the transfer function in the Laplace domain of the PID controller is:

$$L(s) = K_p + K_i / s + K_d s,$$

where s is the complex frequency.

Proportional Term

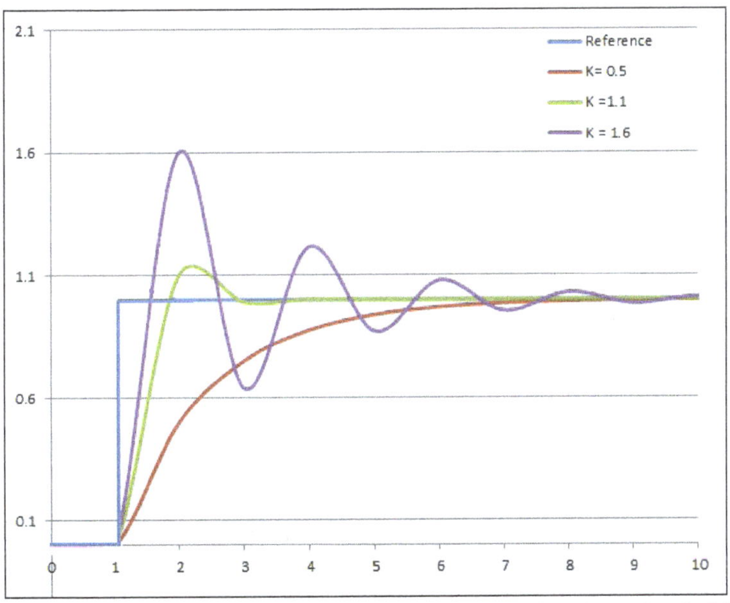

Response of PV to step change of SP vs time, for three values of K_p (K_i and K_d held constant).

The proportional term produces an output value that is proportional to the current error value. The proportional response can be adjusted by multiplying the error by a constant K_p, called the proportional gain constant.

The proportional term is given by,

$$P_{out} = K_p e(t).$$

A high proportional gain results in a large change in the output for a given change in the error. If the proportional gain is too high, the system can become unstable. In contrast, a small gain results in a small output response to a large input error, and a less responsive or less sensitive controller. If the proportional gain is too low, the control action may be too small when responding to system disturbances. Tuning theory and industrial practice indicate that the proportional term should contribute the bulk of the output change.

Steady-state Error

The steady-state error is the difference between the desired final output and the actual one. Because a non-zero error is required to drive it, a proportional controller generally operates with a steady-state error.[a] Steady-state error (SSE) is proportional to the process gain and inversely proportional to proportional gain. SSE may be mitigated by adding a compensating bias term to the setpoint AND output, or corrected dynamically by adding an integral term.

Integral Term

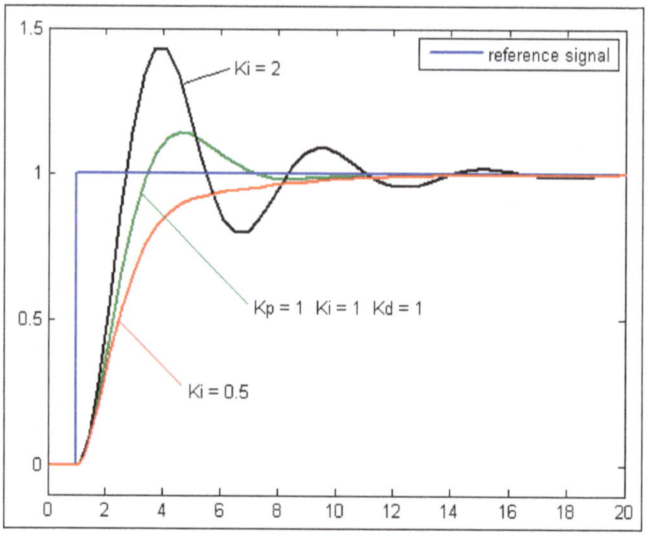

Response of PV to step change of SP vs time, for three values of K_i (K_p and K_d held constant).

The contribution from the integral term is proportional to both the magnitude of the error and the duration of the error. The integral in a PID controller is the sum of the instantaneous error over time and gives the accumulated offset that should have been corrected previously. The accumulated error is then multiplied by the integral gain (K_i) and added to the controller output.

The integral term is given by,

$$I_{out} = K_i \int_0^t e(\tau) d\tau.$$

The integral term accelerates the movement of the process towards setpoint and eliminates the residual steady-state error that occurs with a pure proportional controller. However, since the integral term responds to accumulated errors from the past, it can cause the present value to overshoot the setpoint value.

Derivative Term

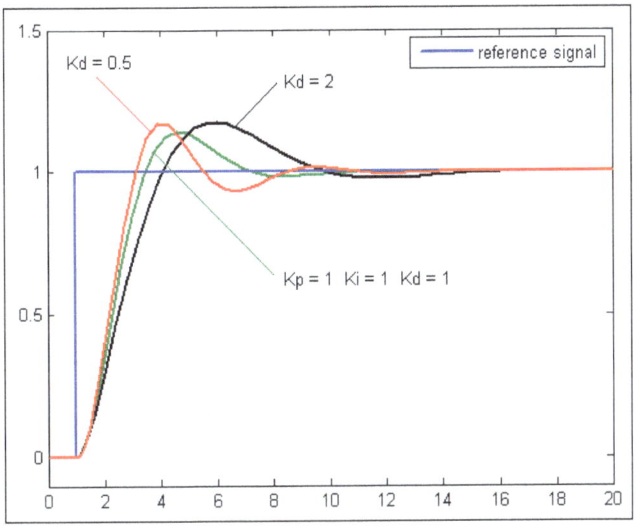

Response of PV to step change of SP vs time, for three values of K_d (K_p and K_i held constant).

The derivative of the process error is calculated by determining the slope of the error over time and multiplying this rate of change by the derivative gain K_d. The magnitude of the contribution of the derivative term to the overall control action is termed the derivative gain, K_d.

The derivative term is given by,

$$D_{out} = K_d \frac{de(t)}{dt}.$$

Derivative action predicts system behavior and thus improves settling time and stability of the system. An ideal derivative is not causal, so that implementations of PID controllers include an additional low-pass filtering for the derivative term to limit the high-frequency gain and noise. Derivative action is seldom used in practice though – by one estimate in only 25% of deployed controllers – because of its variable impact on system stability in real-world applications.

Loop Tuning

Tuning a control loop is the adjustment of its control parameters (proportional band/gain, integral gain/reset, derivative gain/rate) to the optimum values for the desired control response. Stability (no unbounded oscillation) is a basic requirement, but beyond that, different systems have different behavior, different applications have different requirements, and requirements may conflict with one another.

PID tuning is a difficult problem, even though there are only three parameters and in principle is simple to describe, because it must satisfy complex criteria within the limitations of PID control. There are accordingly various methods for loop tuning, and more sophisticated techniques are the subject of patents.

Designing and tuning a PID controller appears to be conceptually intuitive, but can be hard in practice, if multiple (and often conflicting) objectives such as short transient and high stability are to be achieved. PID controllers often provide acceptable control using default tunings, but performance can generally be improved by careful tuning, and performance may be unacceptable with poor tuning. Usually, initial designs need to be adjusted repeatedly through computer simulations until the closed-loop system performs or compromises as desired.

Some processes have a degree of nonlinearity and so parameters that work well at full-load conditions don't work when the process is starting up from no-load; this can be corrected by gain scheduling (using different parameters in different operating regions).

Stability

If the PID controller parameters (the gains of the proportional, integral and derivative terms) are chosen incorrectly, the controlled process input can be unstable, i.e., its output diverges, with or without oscillation, and is limited only by saturation or mechanical breakage. Instability is caused by excess gain, particularly in the presence of significant lag.

Generally, stabilization of response is required and the process must not oscillate for any combination of process conditions and setpoints, though sometimes marginal stability (bounded oscillation) is acceptable or desired.

The total loop transfer function is:

$$H(s) = \frac{K(s)G(s)}{1 + K(s)G(s)}$$

where,

$K(s)$ is the PID transfer function and

$G(s)$ is the plant transfer function.

The system is called unstable where the closed loop transfer function diverges for some s. This happens for situations where $K(s)G(s) = -1$. Typically, this happens when $|K(s)G(s)| = 1$ with a 180 degree phase shift. Stability is guaranteed when $K(s)G(s) < 1$ for frequencies that suffer high phase shifts. A more general formalism of this effect is known as the Nyquist stability criterion.

Optimal Behavior

The optimal behavior on a process change or setpoint change varies depending on the application.

Two basic requirements are regulation (disturbance rejection – staying at a given setpoint) and command tracking (implementing setpoint changes) – these refer to how well the controlled variable tracks the desired value. Specific criteria for command tracking include rise time and settling time. Some processes must not allow an overshoot of the process variable beyond the setpoint if, for example, this would be unsafe. Other processes must minimize the energy expended in reaching a new setpoint.

Tuning Methods

There are several methods for tuning a PID loop. The most effective methods generally involve the development of some form of process model, then choosing P, I, and D based on the dynamic model parameters. Manual tuning methods can be relatively time consuming, particularly for systems with long loop times.

The choice of method will depend largely on whether or not the loop can be taken offline for tuning, and on the response time of the system. If the system can be taken offline, the best tuning method often involves subjecting the system to a step change in input, measuring the output as a function of time, and using this response to determine the control parameters.

Table: Choosing a tuning method.

Method	Advantages	Disadvantages
Manual tuning	No math required; online.	Requires experienced personnel.
Ziegler–Nichols [b]	Proven method; online.	Process upset, some trial-and-error, very aggressive tuning.
Tyreus Luyben	Proven method; online.	Process upset, some trial-and-error, very aggressive tuning.
Software tools	Consistent tuning; online or offline - can employ computer-automated control system design (CAutoD) techniques; may include valve and sensor analysis; allows simulation before downloading; can support non-steady-state (NSS) tuning.	Some cost or training involved.
Cohen–Coon	Good process models.	Some math; offline; only good for first-order processes.
Åström-Hägglund	Can be used for auto tuning; amplitude is minimum so this method has lowest process upset	The process itself is inherently oscillatory.

Manual Tuning

If the system must remain online, one tuning method is to first set K_i and K_d values to zero. Increase the K_p until the output of the loop oscillates, then the K_p should be set to approximately half of that value for a "quarter amplitude decay" type response. Then increase K_i until any offset is corrected in sufficient time for the process. However, too much K_i will cause instability. Finally, increase K_d, if required, until the loop is acceptably quick to reach its reference after a load disturbance. However, too much K_d will cause excessive response and overshoot. A fast PID loop tuning usually overshoots slightly to reach the setpoint more quickly; however, some systems cannot accept overshoot, in which case an overdamped closed-loop system is required, which will require a K_p setting significantly less than half that of the K_p setting that was causing oscillation.

Table: Effects of increasing a parameter independently.

Parameter	Rise time	Overshoot	Settling time	Steady-state error	Stability
K_p	Decrease	Increase	Small change	Decrease	Degrade
K_i	Decrease	Increase	Increase	Eliminate	Degrade
K_d	Minor change	Decrease	Decrease	No effect in theory	Improve if small

Ziegler–Nichols Method

Another heuristic tuning method is formally known as the Ziegler–Nichols method, introduced by John G. Ziegler and Nathaniel B. Nichols in the 1940s. As in the method above, the K_i and K_d gains are first set to zero. The proportional gain is increased until it reaches the ultimate gain, K_u, at which the output of the loop starts to oscillate. K_u and the oscillation period T_u are used to set the gains as follows:

Ziegler–Nichols method			
Control Type	K_p	K_i	K_d
P	$0.50K_u$	–	–
PI	$0.45K_u$	$0.54K_u / T_u$	–
PID	$0.60K_u$	$1.2K_u / T_u$	$3K_uT_u / 40$

These gains apply to the ideal, parallel form of the PID controller. When applied to the standard PID form, only the integral and derivative time parameters T_i and T_d are dependent on the oscillation period T_u.

Cohen–Coon Parameters

This method was developed in 1953 and is based on a first-order + time delay model. Similar to the Ziegler–Nichols method, a set of tuning parameters were developed to yield a closed-loop response with a decay ratio of 1/4. Arguably the biggest problem with these parameters is that a small change in the process parameters could potentially cause a closed-loop system to become unstable.

Relay (Åström–Hägglund) Method

Published in 1984 by Karl Johan Åström and Tore Hägglund, the relay method temporarily operates the process using bang-bang control and measures the resultant oscillations. The output is switched (as if by a relay, hence the name) between two values of the control variable. The values must be chosen so the process will cross the setpoint, but need not be 0% and 100%; by choosing suitable values, dangerous oscillations can be avoided.

As long as the process variable is below the setpoint, the control output is set to the higher value. As soon as it rises above the setpoint, the control output is set to the lower value. Ideally, the output waveform is nearly square, spending equal time above and below the setpoint. The period and amplitude of the resultant oscillations are measured, and used to compute the ultimate gain and period, which are then fed into the Ziegler–Nichols method.

Specifically, the ultimate period T_u is assumed to be equal to the observed period, and the ultimate gain is computed as $K_u = 4b / \pi a$, where a is the amplitude of the process variable oscillation, and b is the amplitude of the control output change which caused it.

There are numerous variants on the relay method.

First Order and Dead Time Model

The transfer function for a first order process, with dead time, is:

$$y(s) = \frac{k_p e^{-\theta s}}{\tau_p s + 1} * u(s)$$

where k_p is the process gain, τ_p is the time constant, θ is the dead time, and u(s) is a step change input. Converting this transfer function to the time domain results in:

$$y(t) = k_p \Delta u(1 - e^{\frac{-t-\theta}{\tau_p}})$$

using the same parameters found above.

It is important when using this method to apply a large enough step change input that the output can be measured; however, too large of a step change can affect the process stability. Additionally, a larger step change will ensure that the output is not changing due to a disturbance (for best results, try to minimize disturbances when performing the step test).

One way to determine the parameters for the first order process is using the 63.2% method. In this method, the process gain (k_p) is equal to the change in output divided by the change in input. The dead time (θ) is the amount of time between when the step change occurred and when the output first changed. The time constant (τ_p) is the amount of time it takes for the output to reach 63.2% of the new steady state value after the step change. One downside to using this method is that the time to reach a new steady state value can take a while if the process has a large time constants.

PID Tuning Software

Most modern industrial facilities no longer tune loops using the manual calculation methods. Instead, PID tuning and loop optimization software are used to ensure consistent results. These software packages will gather the data, develop process models, and suggest optimal tuning. Some software packages can even develop tuning by gathering data from reference changes.

Mathematical PID loop tuning induces an impulse in the system, and then uses the controlled system's frequency response to design the PID loop values. In loops with response times of several minutes, mathematical loop tuning is recommended, because trial and error can take days just to find a stable set of loop values. Optimal values are harder to find. Some digital loop controllers offer a self-tuning feature in which very small setpoint changes are sent to the process, allowing the controller itself to calculate optimal tuning values.

Another approach calculates initial values via the Ziegler–Nichols method, and uses a numerical optimization technique to find better PID coefficients. Other formulas are available to tune the loop according to different performance criteria. Many patented formulas are now embedded within PID tuning software and hardware modules.

Advances in automated PID loop tuning software also deliver algorithms for tuning PID Loops in a dynamic or non-steady state (NSS) scenario. The software will model the dynamics of a process, through a disturbance, and calculate PID control parameters in response.

Limitations of PID Control

While PID controllers are applicable to many control problems, and often perform satisfactorily without any improvements or only coarse tuning, they can perform poorly in some applications, and do not in general provide optimal control. The fundamental difficulty with PID control is that it is a feedback control system, with constant parameters, and no direct knowledge of the process, and thus overall performance is reactive and a compromise. While PID control is the best controller in an observer without a model of the process, better performance can be obtained by overtly modeling the actor of the process without resorting to an observer.

PID controllers, when used alone, can give poor performance when the PID loop gains must be reduced so that the control system does not overshoot, oscillate or hunt about the control setpoint value. They also have difficulties in the presence of non-linearities, may trade-off regulation versus response time, do not react to changing process behavior (say, the process changes after it has warmed up), and have lag in responding to large disturbances.

The most significant improvement is to incorporate feed-forward control with knowledge about the system, and using the PID only to control error. Alternatively, PIDs can be modified in more minor ways, such as by changing the parameters (either gain scheduling in different use cases or adaptively modifying them based on performance), improving measurement (higher sampling rate, precision, and accuracy, and low-pass filtering if necessary), or cascading multiple PID controllers.

Linearity

Another problem faced with PID controllers is that they are linear, and in particular symmetric. Thus, performance of PID controllers in non-linear systems (such as HVAC systems) is variable. For example, in temperature control, a common use case is active heating (via a heating element) but passive cooling (heating off, but no cooling), so overshoot can only be corrected slowly – it cannot be forced downward. In this case the PID should be tuned to be overdamped, to prevent or reduce overshoot, though this reduces performance (it increases settling time).

Noise in Derivative

A problem with the derivative term is that it amplifies higher frequency measurement or process noise that can cause large amounts of change in the output. It is often helpful to filter the measurements with a low-pass filter in order to remove higher-frequency noise components. As low-pass filtering and derivative control can cancel each other out, the amount of filtering is limited. Therefore, low noise instrumentation can be important. A nonlinear median filter may be used, which improves the filtering efficiency and practical performance. In some cases, the differential band can be turned off with little loss of control. This is equivalent to using the PID controller as a PI controller.

Modifications to the PID Algorithm

The basic PID algorithm presents some challenges in control applications that have been addressed by minor modifications to the PID form.

Integral Windup

One common problem resulting from the ideal PID implementations is integral windup. Following a large change in setpoint the integral term can accumulate an error larger than the maximal value for the regulation variable (windup), thus the system overshoots and continues to increase until this accumulated error is unwound. This problem can be addressed by:

- Disabling the integration until the PV has entered the controllable region.
- Preventing the integral term from accumulating above or below pre-determined bounds.
- Back-calculating the integral term to constrain the regulator output within feasible bounds.

Overshooting from Known Disturbances

For example, a PID loop is used to control the temperature of an electric resistance furnace where the system has stabilized. Now when the door is opened and something cold is put into the furnace the temperature drops below the setpoint. The integral function of the controller tends to compensate for error by introducing another error in the positive direction. This overshoot can be avoided by freezing of the integral function after the opening of the door for the time the control loop typically needs to reheat the furnace.

PI Controller

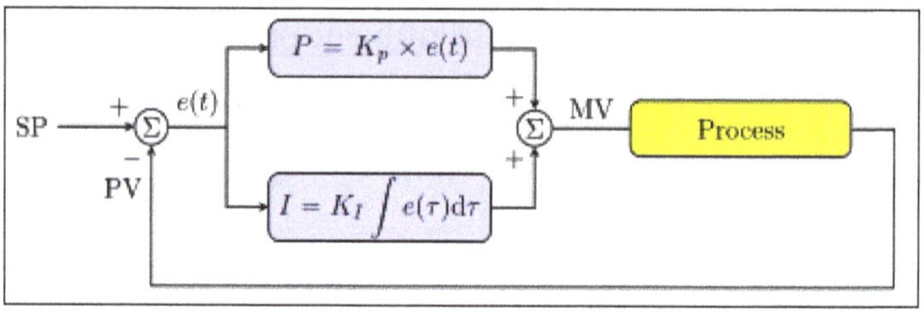

Basic block of a PI controller.

A PI controller (proportional-integral controller) is a special case of the PID controller in which the derivative (D) of the error is not used.

The controller output is given by,

$$K_P\Delta + K_I\int \Delta dt$$

where Δ is the error or deviation of actual measured value (PV) from the setpoint (SP).

$$\Delta = SP - PV.$$

A PI controller can be modelled easily in software such as Simulink or Xcos using a "flow chart" box involving Laplace operators:

$$C = \frac{G(1+\tau s)}{\tau s}$$

where,

$$G = K_P = \text{proportional gain}$$

$$\frac{G}{\tau} = K_I = \text{integral gain}$$

Setting a value for G is often a tradeoff between decreasing overshoot and increasing settling time.

The lack of derivative action may make the system more steady in the steady state in the case of noisy data. This is because derivative action is more sensitive to higher-frequency terms in the inputs.

Without derivative action, a PI-controlled system is less responsive to real (non-noise) and relatively fast alterations in state and so the system will be slower to reach setpoint and slower to respond to perturbations than a well-tuned PID system may be.

Deadband

Many PID loops control a mechanical device (for example, a valve). Mechanical maintenance can be a major cost and wear leads to control degradation in the form of either stiction or backlash in the mechanical response to an input signal. The rate of mechanical wear is mainly a function of how often a device is activated to make a change. Where wear is a significant concern, the PID loop may have an output deadband to reduce the frequency of activation of the output (valve). This is accomplished by modifying the controller to hold its output steady if the change would be small (within the defined deadband range). The calculated output must leave the deadband before the actual output will change.

Setpoint Step Change

The proportional and derivative terms can produce excessive movement in the output when a system is subjected to an instantaneous step increase in the error, such as a large setpoint change. In the case of the derivative term, this is due to taking the derivative of the error, which is very large in the case of an instantaneous step change. As a result, some PID algorithms incorporate some of the following modifications:

- Setpoint ramping: In this modification, the setpoint is gradually moved from its old value to a newly specified value using a linear or first order differential ramp function. This avoids the discontinuity present in a simple step change.
- Derivative of the process variable: In this case the PID controller measures the derivative of the measured process variable (PV), rather than the derivative of the error. This quantity is always continuous (i.e., never has a step change as a result of changed setpoint). This modification is a simple case of setpoint weighting.
- Setpoint weighting: Setpoint weighting adds adjustable factors (usually between 0 and 1) to the setpoint in the error in the proportional and derivative element

of the controller. The error in the integral term must be the true control error to avoid steady-state control errors. These two extra parameters do not affect the response to load disturbances and measurement noise and can be tuned to improve the controller's setpoint response.

Feed-forward

The control system performance can be improved by combining the feedback (or closed-loop) control of a PID controller with feed-forward (or open-loop) control. Knowledge about the system (such as the desired acceleration and inertia) can be fed forward and combined with the PID output to improve the overall system performance. The feed-forward value alone can often provide the major portion of the controller output. The PID controller primarily has to compensate whatever difference or error remains between the setpoint (SP) and the system response to the open loop control. Since the feed-forward output is not affected by the process feedback, it can never cause the control system to oscillate, thus improving the system response without affecting stability. Feed forward can be based on the setpoint and on extra measured disturbances. Setpoint weighting is a simple form of feed forward.

For example, in most motion control systems, in order to accelerate a mechanical load under control, more force is required from the actuator. If a velocity loop PID controller is being used to control the speed of the load and command the force being applied by the actuator, then it is beneficial to take the desired instantaneous acceleration, scale that value appropriately and add it to the output of the PID velocity loop controller. This means that whenever the load is being accelerated or decelerated, a proportional amount of force is commanded from the actuator regardless of the feedback value. The PID loop in this situation uses the feedback information to change the combined output to reduce the remaining difference between the process setpoint and the feedback value. Working together, the combined open-loop feed-forward controller and closed-loop PID controller can provide a more responsive control system.

Bumpless Operation

PID controllers are often implemented with a "bumpless" initialization feature that recalculates the integral accumulator term to maintain a consistent process output through parameter changes. A partial implementation is to store the integral of the integral gain times the error rather than storing the integral of the error and postmultiplying by the integral gain, which prevents discontinuous output when the I gain is changed, but not the P or D gains.

Other Improvements

In addition to feed-forward, PID controllers are often enhanced through methods such as PID gain scheduling (changing parameters in different operating conditions), fuzzy logic, or computational verb logic. Further practical application issues can arise from instrumentation connected to the controller. A high enough sampling rate, measurement precision, and measurement accuracy are required to achieve adequate control performance. Another new method for improvement of PID controller is to increase the degree of freedom by using fractional order. The order of the integrator and differentiator add increased flexibility to the controller.

Cascade Control

One distinctive advantage of PID controllers is that two PID controllers can be used together to yield better dynamic performance. This is called cascaded PID control. In cascade control there are two PIDs arranged with one PID controlling the setpoint of another. A PID controller acts as outer loop controller, which controls the primary physical parameter, such as fluid level or velocity. The other controller acts as inner loop controller, which reads the output of outer loop controller as setpoint, usually controlling a more rapid changing parameter, flowrate or acceleration. It can be mathematically proven that the working frequency of the controller is increased and the time constant of the object is reduced by using cascaded PID controllers.

For example, a temperature-controlled circulating bath has two PID controllers in cascade, each with its own thermocouple temperature sensor. The outer controller controls the temperature of the water using a thermocouple located far from the heater, where it accurately reads the temperature of the bulk of the water. The error term of this PID controller is the difference between the desired bath temperature and measured temperature. Instead of controlling the heater directly, the outer PID controller sets a heater temperature goal for the inner PID controller. The inner PID controller controls the temperature of the heater using a thermocouple attached to the heater. The inner controller's error term is the difference between this heater temperature setpoint and the measured temperature of the heater. Its output controls the actual heater to stay near this setpoint.

The proportional, integral, and differential terms of the two controllers will be very different. The outer PID controller has a long time constant – all the water in the tank needs to heat up or cool down. The inner loop responds much more quickly. Each controller can be tuned to match the physics of the system *it* controls – heat transfer and thermal mass of the whole tank or of just the heater – giving better total response.

Alternative Nomenclature and PID Forms

Standard versus Parallel (ideal) PID Form

The form of the PID controller most often encountered in industry, and the one most relevant to tuning algorithms is the standard form. In this form the K_p gain is applied to the I_{out}, and D_{out} terms, yielding:

$$u(t) = K_p\left(e(t) + \frac{1}{T_i}\int_0^t e(\tau)d\tau + T_d\frac{d}{dt}e(t)\right)$$

where,

T_i is the integral time.

T_d is the derivative time.

In this standard form, the parameters have a clear physical meaning. In particular, the inner summation produces a new single error value which is compensated for future and past errors. The proportional error term is the current error. The derivative components term attempts to predict the error value at T_d seconds (or samples) in the future, assuming that the loop control remains unchanged. The integral component adjusts the error value to compensate for the sum of all past errors, with the intention of completely eliminating them in T_i seconds (or samples). The resulting compensated single error value is then scaled by the single gain K_p to compute the control variable.

In the parallel form, shown in the controller theory section,

$$u(t) = K_p e(t) + K_i\int_0^t e(\tau)d\tau + K_d\frac{d}{dt}e(t)$$

the gain parameters are related to the parameters of the standard form through $K_i = K_p / T_i$ and $K_d = K_p T_d$. This parallel form, where the parameters are treated as simple gains, is the most general and flexible form. However, it is also the form where the parameters have the least physical interpretation and is generally reserved for theoretical treatment of the PID controller. The standard form, despite being slightly more complex mathematically, is more common in industry.

Reciprocal Gain

In many cases, the manipulated variable output by the PID controller is a dimensionless fraction between 0 and 100% of some maximum possible value, and the translation into real units (such as pumping rate or watts of heater power) is outside the PID controller. The process variable, however, is in dimensioned units such as temperature. It is common in this case to express the gain K_p not as "output per degree", but rather in the reciprocal form of a proportional band $100 / K_p$, which is "degrees per full output": the range over which the output changes from 0 to 1 (0% to 100%). Beyond this range, the output is saturated, full-off or full-on. The narrower this band, the higher the proportional gain.

Basing Derivative Action on PV

In most commercial control systems, derivative action is based on process variable rather than error. That is, a change in the setpoint does not affect the derivative action. This is because the digitized version of the algorithm produces a large unwanted spike when the setpoint is changed. If the setpoint is constant then changes in the PV will be the same as changes in error. Therefore, this modification makes no difference to the way the controller responds to process disturbances.

Basing Proportional Action on PV

Most commercial control systems offer the option of also basing the proportional action solely on the process variable. This means that only the integral action responds to changes in the setpoint. The modification to the algorithm does not affect the way the controller responds to process disturbances. Basing proportional action on PV eliminates the instant and possibly very large change in output caused by a sudden change to the setpoint. Depending on the process and tuning this may be beneficial to the response to a setpoint step.

$$MV(t) = K_p \left(-PV(t) + \frac{1}{T_i} \int_o^t e(\tau) d\tau - T_d \frac{d}{dt} PV(t) \right)$$

Laplace Form of the PID Controller

Sometimes it is useful to write the PID regulator in Laplace transform form:

$$G(s) = K_p + \frac{K_i}{s} + K_d s = \frac{K_d s^2 + K_p s + K_i}{s}$$

Having the PID controller written in Laplace form and having the transfer function of the controlled system makes it easy to determine the closed-loop transfer function of the system.

Series/Interacting Form

Another representation of the PID controller is the series, or interacting form:

$$G(s) = K_c \frac{(\tau_i s + 1)}{\tau_i s} (\tau_d s + 1)$$

where the parameters are related to the parameters of the standard form through,

$$K_p = K_c \cdot \alpha,$$

$$T_i = \tau_i \cdot \alpha$$

and,

$$T_d = \frac{\tau_d}{\alpha}$$

with,

$$\alpha = 1 + \frac{\tau_d}{\tau_i}.$$

This form essentially consists of a PD and PI controller in series, and it made early (analog) controllers easier to build. When the controllers later became digital, many kept using the interacting form.

Discrete Implementation

The analysis for designing a digital implementation of a PID controller in a microcontroller (MCU) or FPGA device requires the standard form of the PID controller to be discretized. Approximations for first-order derivatives are made by backward finite differences. The integral term is discretized, with a sampling time Δt, as follows:

$$\int_o^{t_k} e(\tau)d\tau = \sum_{i=1}^{k} e(t_i)\Delta t$$

The derivative term is approximated as,

$$\frac{de(t_k)}{dt} = \frac{e(t_k) - e(t_{k-1})}{\Delta t}$$

Thus, a velocity algorithm for implementation of the discretized PID controller in a MCU is obtained by differentiating $u(t)$, using the numerical definitions of the first and second derivative and solving for $u(t_k)$ and finally obtaining:

$$u(t_k) = u(t_{k-1}) + K_p\left[\left(1 + \frac{\Delta t}{T_i} + \frac{T_d}{\Delta t}\right)e(t_k) + \left(-1 - \frac{2T_d}{\Delta t}\right)e(t_{k-1}) + \frac{T_d}{\Delta t}e(t_{k-2})\right]$$

$$\text{s.t. } T_i = K_p / K_i, T_d = K_d / K_p.$$

Pseudocode

Here is a simple software loop that implements a PID algorithm:

```
previous_error := 0
```

```
integral := 0
loop:
    error := setpoint - measured_value
    integral := integral + error × dt
    derivative := (error - previous_error) /dt
    output := Kp × error + Ki × integral + Kd × derivative
    previous_error := error
    wait(dt)
    goto loop
```

In this example, two variables that will be maintained within the loop are initialized to zero, then the loop begins. The current error is calculated by subtracting the measured_ value (the process variable, or PV) from the current setpoint (SP). Then, integral and derivative values are calculated, and these and the error are combined with three preset gain terms – the proportional gain, the integral gain and the derivative gain – to derive an output value.

In the real world, this is D-to-A converted and passed into the process under control as the manipulated variable (MV). The current error is stored elsewhere for re-use in the next differentiation, the program then waits until dt seconds have passed since start, and the loop begins again, reading in new values for the PV and the setpoint and calculating a new value for the error. Note that for real code, the use of "wait(dt)" might be inappropriate because it doesn't account for time taken by the algorithm itself during the loop, or more importantly, any preemption delaying the algorithm.

Sequential Control

A control system in which the individual steps are processed in a predetermined order, progression from one sequence step to the next being dependent on defined conditions being satisfied. Such a system may be time-dependent, in which the step transition conditions are functions of time only; on external-event dependent, where the conditions are functions of Input signals or combinations of these conditions.

Object has finite numbers of states: lift, Clean-in-Place (CIP) processes (is a method of cleaning the interior surfaces of pipes, vessels, process equipment, filters and associated fittings, without disassembly), assembly line process, conveyor systems, robots, power protection systems, motor starting and control.

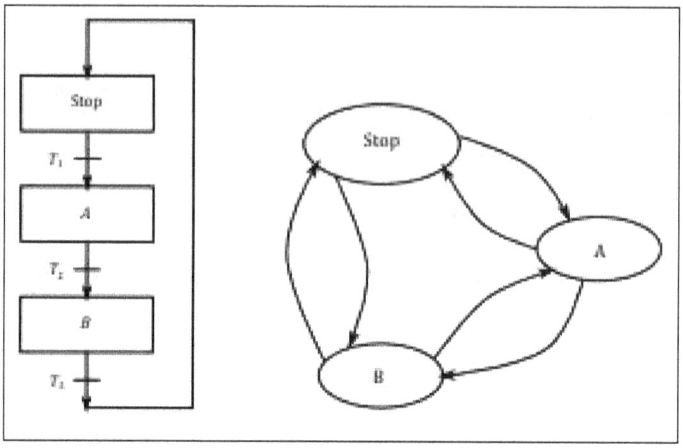

State and Data-flow diagrams. State diagram: states (initial state), transactions, conditions.

Table: Comparison of the Control Techniques.

Process control	Sequential
Feedback is continuous.	Status I/O typically bi-level in nature.
Controller could be implemented with analog or digital control methods.	ON-OFF control.

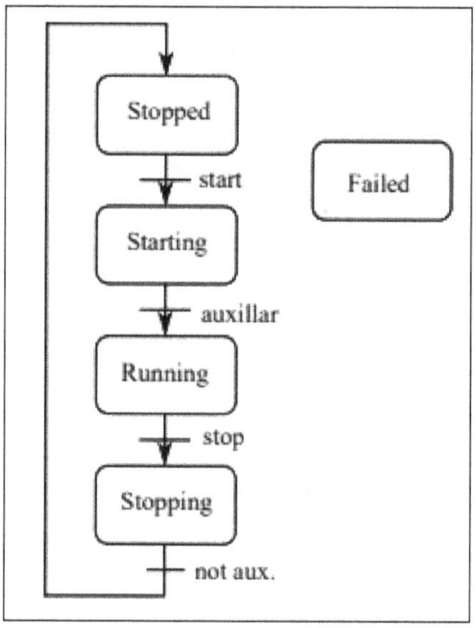

State based control - states are fixed and clear transactions.

Possible states of the controlled process/equipment:

- Final states: Stopped, running.
- Transition states: Starting, stopping.
- Failure states.
- Intermediate states: Waiting, maintenance, cleaning, loading, etc.

Structure

- Handler: The execution engine for a Sequential Control Module (SCM). There are Main handlers and Exception handlers.
- Step block: Executes output expressions that change process parameters and provide instructions to the operator.
- Transition block: Waits for specific process conditions to be true. When the logical statement formed by a transition's conditions and logic gates is true, the SCM executes the next step or synchronization block.
- Synchronization block: Enables the SCM to execute parallel paths contained within the handler.

Structure of the program - steps: A, B, C. Each state is either active or passive. Only one state is active at a time. The state machine always starts in a particular state defined as the initial state, and it ends (stops) after the final state. Special symbols or colors may be used for the initial state and the final state.

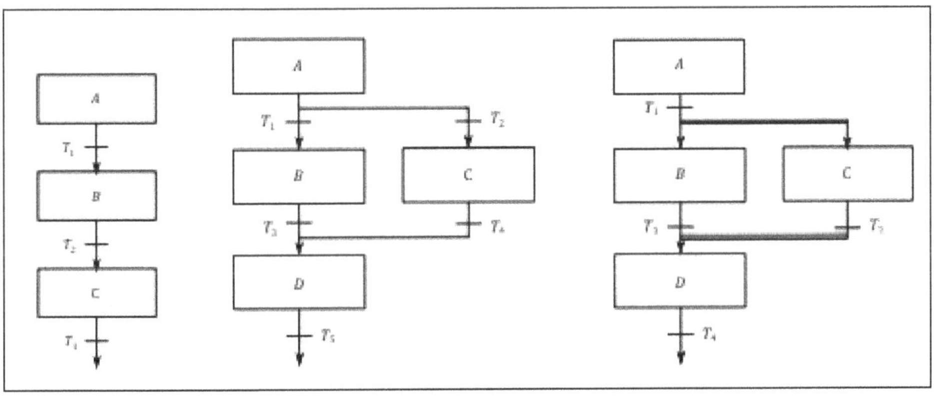

Sequence types: Sequence, Choice and Parallel actions.

If no interrupt or abnormal conditions occur, a handler executes transitions, steps, and phases in sequential order until reaching the end of the module's program where the module is either completed or returned to start. Execution can be interrupted by predetermined conditions that can interrupt, stop, hold, or abort the SCM.

Every SCM must have at least one main handler, but it can have multiple main handlers and optional exception handlers. While only one handler can be actively executing, exception handlers can be waiting for specified process conditions. When the specified process condition occurs, the Main Handler's process actions cease execution and the exception handler (Abort, Stop, Hold, and Interrupt) begins its ordered set of process actions.

A Handler must have exactly one Invoke Transition which allows to specify logical conditions for invocation of the handler and specifies the first step of the handler.

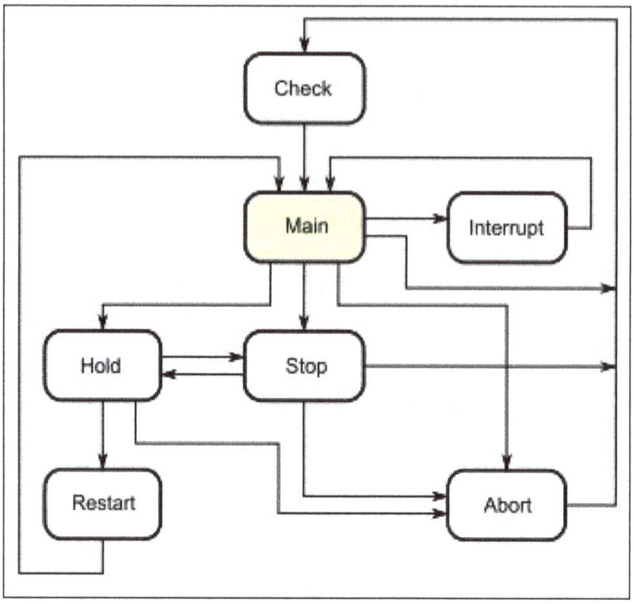

Configurable Handlers categories.

Handlers are prioritized from Main to Abort. Main is the lowest priority; Abort has the highest priority. If the Main handler is active and the Hold conditions become true, the Hold handler becomes the active handler. If at the same instance, conditions for the Abort handler become true, the Abort handler becomes the active handler. Main Handler contains the primary sequential activities of the process. Multiple Main handlers may be configured, but only one will be selected for execution at any given time. Check Handler is a normal exception handler which can be configured to initialize process equipment and reset values in order to setup and verify the base conditions required for the SCM to execute properly. No other handler is enabled when the Check Handler is executing.

Example: Two positional level control.

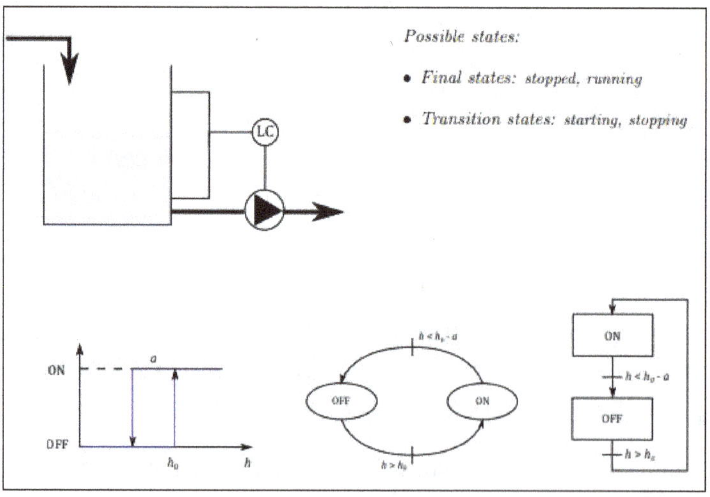

Pump On-OFF Control.

Example: Two positional level control with additional pump.

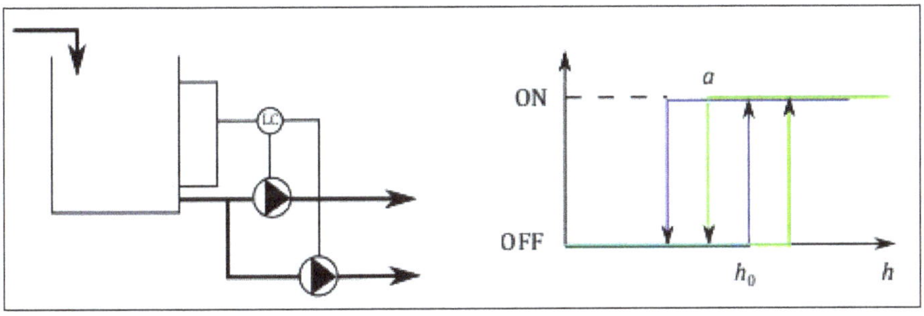

Two pumps control.

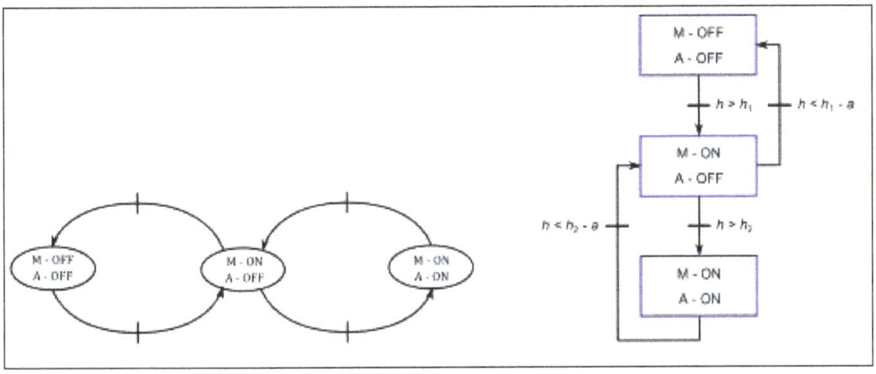

Control diagrams.

Batch Control

Producc: with thc required quantity, quality, safely, within a short period, min. pollution, etc. Production is either continuous (electricity), discrete (car), or batch. Earlier any production took place by portions and was a handmade.

- Brewing /4000 years old!/,
- Cooking,
- (gun)powder, steel, soap, etc.

Modern examples of batch processes are:

- Washing machine work, concrete, bread making.
- Varnishes, paints, biotechnology, pharmaceutical.

Each process has a beginning, end (start, stop) and a life-cycle. The batch production allows you to:

- Change the production number, characteristics, qualities.
- Process can be monitored, studied, corrected, repeated, etc.

Features of the batch processes:

- The volume is limited and varies.
- Loading components are kept with ratio A:B:C.
- Controlled from beginning to the end, changes are large, complex.
- Process dynamics depends on the quantity, components, here is no equilibrium point (SP 6=const), limitations of signals, time constants can change (up to ×10), nonlinearity.
- Control of the completion (quality or composition of the product).
- Preparation errors can cause damage of the product or devices.

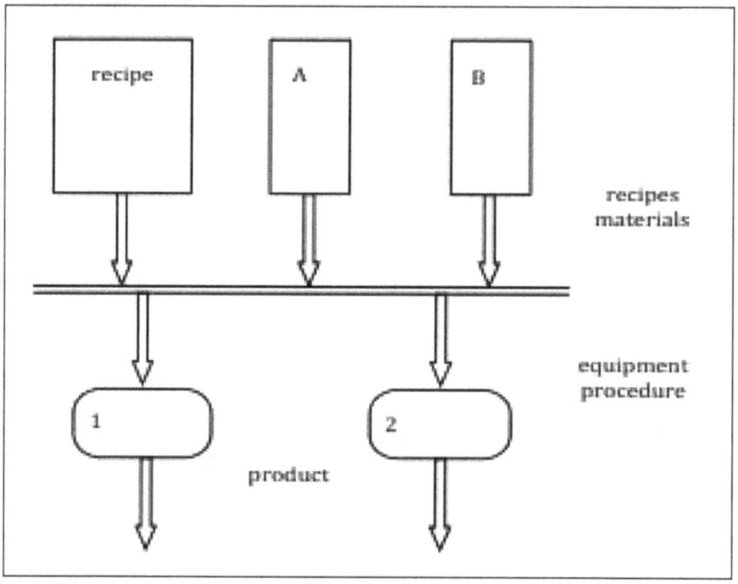

Batch process.

With batch processes a certain amount of product is produced:

- Specified material (recipe).
- Fixed sequence of operations (procedure).
- With one or several devices.

Batch processing has its own standards:

- ISA standard S88.
- IEC61512.
- The product is separated from the devices.

- Various products manufactured on the same equipment.
- Changes in the product does not affect the other products.

Recipe

The layered recipe function lets you build recipes in layers to support the S88 recipe hierarchy. In this hierarchy, a higher level recipe can control its underlying recipe(s). Recipes at each layer are implemented as a modular function block. Recipe Control Module (RCM) blocks can represent Procedures, Unit Procedures, and Operations.

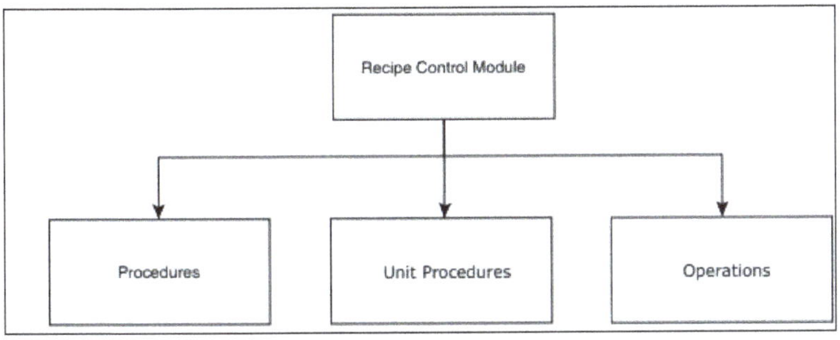

RCM.

Sequential Control Module (SCM) blocks can represent Phases. Phase blocks allow to link layers together to form layered recipes. Phases are implemented as SCMs, which coordinate device control.

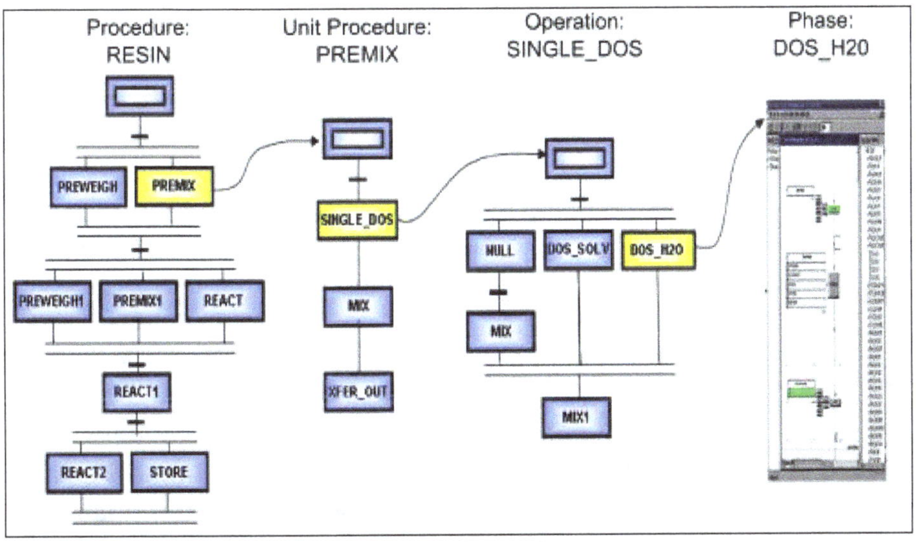

S88 recipe hierarchy.

The RCMs and SCMs at each layer manage their own execution. At run time, the upper level module commands its child (children) through Phase block(s). Phase block monitors the execution of its linked RCM/SCM and projects the key parameters, State, Execution Status, Mode, and Mode Attribute to the parent recipe.

Unit Control Module

The Unit Control Module (UCM) is a container that represents a piece of or logical grouping of physical equipment. The UCM Identifies the current running RCM. Multiple RCMs can be queued to a single UCM, but only one RCM can run against an UCM at a time. UCM does not have a containment relationship with the RCMs that are associated with it.

The UCM maintains an arbitration list that holds the tag names of all the RCMs that are currently waiting for the UCM for their procedure execution. When the current RCM is completed, the first one on the list will be started automatically.

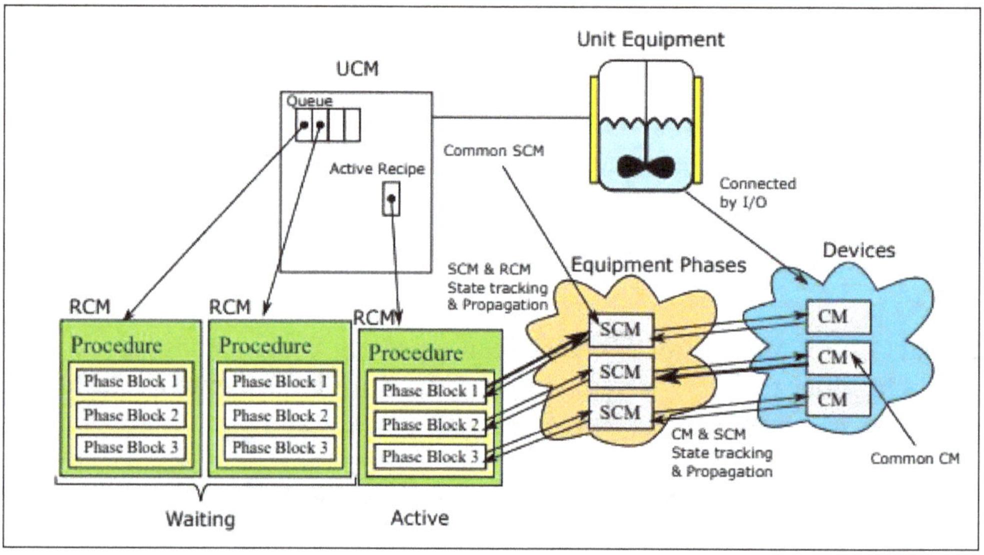

Relationship between UCM, Equipment, and RCM.

A system without resource conflicts (all requesters can get their resources immediately all the time) will work the same all the time (no-conflict system).

Motion Control

Motion control is a sub-field of automation, encompassing the systems or sub-systems involved in moving parts of machines in a controlled manner. The main components involved typically include a motion controller, an energy amplifier, and one or more prime movers or actuators. Motion control may be open loop or closed loop. In open loop systems, the controller sends a command through the amplifier to the prime mover or actuator, and does not know if the desired motion was actually achieved. Typical systems include stepper motor or fan control. For tighter control with more precision, a measuring device may be added to the system (usually near the end motion). When the measurement is converted to a signal that is sent back to the controller, and the controller compensates for any error, it becomes a Closed loop System.

Typically the position or velocity of machines are controlled using some type of device such as a hydraulic pump, linear actuator, or electric motor, generally a servo. Motion control is an important part of robotics and CNC machine tools, however in these instances it is more complex than when used with specialized machines, where the kinematics are usually simpler. The latter is often called General Motion Control (GMC). Motion control is widely used in the packaging, printing, textile, semiconductor production, and assembly industries. Motion Control encompasses every technology related to the movement of objects. It covers every motion system from micro-sized systems such as silicon-type micro induction actuators to micro-siml systems such as a space platform. But, these days, the focus of motion control is the special control technology of motion systems with electric actuators such as dc/ac servo motors. Control of robotic manipulators is also included in the field of motion control because most of robotic manipulators are driven by electrical servo motors and the key objective is the control of motion.

The basic architecture of a motion control system contains:

- A motion controller, which calculates and controls the mechanical trajectories (motion profile) an actuator must follow (*i.e.*, motion planning) and, in closed loop systems, employs feedback to make control corrections and thus implement closed-loop control.
- A drive or amplifier to transform the control signal from the motion controller into energy that is presented to the actuator. Newer "intelligent" drives can close the position and velocity loops internally, resulting in much more accurate control.
- A prime mover or actuator such as a hydraulic pump, pneumatic cylinder, linear actuator, or electric motor for output motion.
- In closed loop systems, one or more feedback sensors such as absolute and incremental encoders, resolvers or Hall effect devices to return the position or velocity of the actuator to the motion controller in order to close the position or velocity control loops.
- Mechanical components to transform the motion of the actuator into the desired motion, including: gears, shafting, ball screw, belts, linkages, and linear and rotational bearings.

The interface between the motion controller and drives it control is very critical when coordinated motion is required, as it must provide tight synchronization. Historically the only open interface was an analog signal, until open interfaces were developed that satisfied the requirements of coordinated motion control, the first being SERCOS in 1991 which is now enhanced to SERCOS III. Later interfaces capable of motion control include Ethernet/IP, Profinet IRT, Ethernet Powerlink, and EtherCAT.

Common control functions include:

- Velocity control.
- Position (point-to-point) control: There are several methods for computing a motion trajectory. These are often based on the velocity profiles of a move such as a triangular profile, trapezoidal profile, or an S-curve profile.
- Pressure or Force control.
- Impedance control: This type of control is suitable for environment interaction and object manipulation, such as in robotics.
- Electronic gearing (or cam profiling): The position of a slave axis is mathematically linked to the position of a master axis. A good example of this would be in a system where two rotating drums turn at a given ratio to each other. A more advanced case of electronic gearing is electronic camming. With electronic camming, a slave axis follows a profile that is a function of the master position. This profile need not be salted, but it must be an animated function.

Adjustable-speed Drive

Adjustable speed drive (ASD), also known as variable-speed drive (VSD), describes equipment used to control the speed of machinery. Many industrial processes such as assembly lines must operate at different speeds for different products. Where process conditions demand adjustment of flow from a pump or fan, varying the speed of the drive may save energy compared with other techniques for flow control.

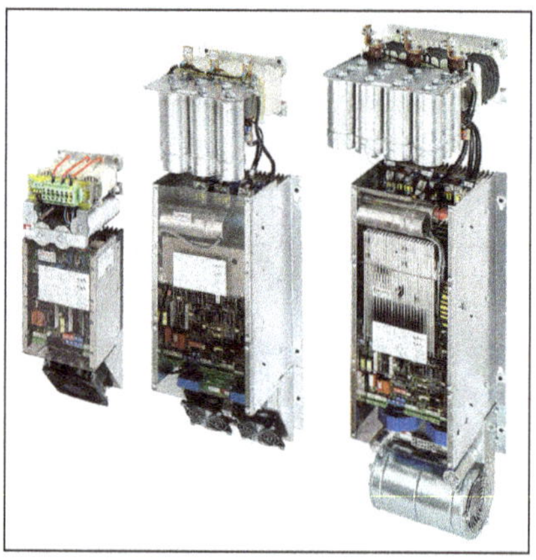

Line regenerative variable frequency drives, showing capacitors (top cylinders) and inductors attached which filter the regenerated power.

Where speeds may be selected from several different pre-set ranges, usually the drive is said to be adjustable speed. If the output speed can be changed without steps over a range, the drive is usually referred to as variable speed.

Adjustable and variable speed drives may be purely mechanical (termed variators), electromechanical, hydraulic, or electronic.

Electric Motors

AC electric motors can be run in fixed-speed operation determined by the number of stator pole pairs in the motor and the frequency of the alternating current supply. AC motors can be made for "pole changing" operation, reconnecting the stator winding to vary the number of poles so that two, sometimes three, speeds are obtained. For example a machine with 8 physical pairs of poles, could be connected to allow running with either 4 or 8 pole pairs, giving two speeds - at 60 Hz, these would be 1800 RPM and 900 RPM. If speed changes are rare, the motor may be initially connected for one speed then re-wired for the other speed as process conditions change, or, magnetic contactors can be used to switch between the two speeds as process needs fluctuate. Connections for more than three speeds are uneconomic.

The number of such fixed-speed-operation speeds is constrained by cost as number of pole pairs increases. If many different speeds or continuously variable speeds are required, other methods are required. Direct-current motors allow for changes of speed by adjusting the shunt field current. Another way of changing speed of a direct current motor is to change the voltage applied to the armature.

An adjustable speed drive might consist of an electric motor and controller that is used to adjust the motor's operating speed. The combination of a constant-speed motor and a continuously adjustable mechanical speed-changing device might also be called an adjustable speed drive. Power electronics based variable frequency drives are rapidly making older technology redundant.

Reasons for using Adjustable Speed Drives

Process control and energy conservation are the two primary reasons for using an adjustable speed drive. Historically, adjustable speed drives were developed for process control, but energy conservation has emerged as an equally important objective.

Acceleration Control

An adjustable speed drive can often provide smoother operation compared to an alternative fixed speed mode of operation. For example, in a sewage lift station sewage usually flows through sewer pipes under the force of gravity to a wet well location. From there it is pumped up to a treatment process. When fixed speed pumps are used, the pumps are set to start when the level of the liquid in the wet well reaches some high point and stop when the level has been reduced to a low point. Cycling the pumps on and off results in frequent high surges of electric current to start the motors that results in electromagnetic and thermal stresses in the motors and power control equipment, the pumps and pipes are subjected to mechanical and hydraulic stresses, and the sewage treatment process is forced to accommodate surges in the flow of sewage through the process. When adjustable speed drives are used, the pumps operate continuously at a speed that increases as the wet well level increases. This matches the outflow to the average inflow and provides a much smoother operation of the process.

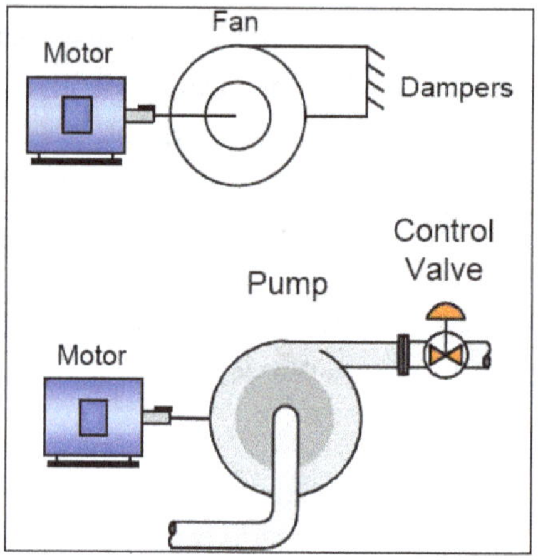

Saving Energy by using Efficient Adjustable Speed Drives

Fans and pumps consume a large part of the energy used by industrial electrical motors. Where fans and pumps serve a varying process load, a simple way to vary the delivered quantity of fluid is with a damper or valve in the outlet of the fan or pump, which by its increased pressure drop, reduces the flow in the process. However, this additional pressure drop represents energy loss. Sometimes it is economically practical to put in some device that recovers this otherwise lost energy. With a variable speed drive on the pump or fan, the supply can be adjusted to match demand and no extra loss is introduced.

For example, when a fan is driven directly by a fixed-speed motor, the airflow is designed for the maximum demand of the system, and so will usually be higher than it needs to be. Airflow can be regulated using a damper but it is more efficient to directly regulate fan motor speed. Following the affinity laws, for 50% of the airflow, the variable-speed motor consumes about 20% of the input power (Amps). The fixed-speed motor still consumes about 85% of the input power at half the flow.

Types of Adjustable Speed Drives

Some prime movers (internal combustion engines, reciprocating or turbine steam engines, water wheels, and others) have a range of operating speeds which can be varied continuously (by adjusting fuel rate or similar means). However, efficiency may be low at extremes of the speed range, and there may be system reasons why the prime mover speed cannot be maintained at very low or very high speeds.

Before electric motors were invented, mechanical speed changers were used to control the mechanical power provided by water wheels and steam engines. When electric motors came into use, means of controlling their speed were developed almost immediately. Today, various types of mechanical drives, hydraulic drives and electric drives compete with one another in the industrial drives market.

Mechanical Adjustable Speed Drives

There are two types of mechanical drives, variable pitch drives and traction drives:

- Variable pitch drives are pulley and belt drives in which the pitch diameter of one or both pulleys can be adjusted.
- Traction drives transmit power through metal rollers running against mating metal rollers. The input/output speed ratio is adjusted by moving the rollers to change the diameters of the contact path. Many different roller shapes and mechanical designs have been used.

Hydraulic Adjustable Speed Drives

There are three types of hydraulic drives, those are: hydrostatic drives, hydrodynamic drives and hydroviscous drives.

- A hydrostatic drive consists of a hydraulic pump and a hydraulic motor. Since positive displacement pumps and motors are used, one revolution of the pump or motor corresponds to a set volume of fluid flow that is determined by the displacement regardless of speed or torque. Speed is regulated by regulating the fluid flow with a valve or by changing the displacement of the pump or motor. Many different design variations have been used. A swash plate drive employs an axial piston pump and motor in which the swash plate angle can be changed to adjust the displacement and thus adjust the speed.
- Hydrodynamic drives or fluid couplings use oil to transmit torque between an impeller on the constant-speed input shaft and a rotor on the adjustable-speed output shaft. The torque converter in the automatic transmission of a car is a hydrodynamic drive.
- A hydroviscous drive consists of one or more discs connected to the input shaft pressed against a similar disc or discs connected to the output shaft. Torque is transmitted from the input shaft to the output shaft through an oil film between the discs. The transmitted torque is proportional to the pressure exerted by a hydraulic cylinder that presses the discs together. This effect may be used as a clutch, such at the Hele-Shaw clutch, or as a variable-speed drive, such as the Beier variable-ratio gear.

Continuously Variable Transmission (CVT)

Mechanical and hydraulic adjustable speed drives are usually called transmissions or continuously variable transmissions when they are used in vehicles, farm equipment and some other types of equipment.

Electric Adjustable Speed Drives

Types of Control

Control can mean either manually adjustable - by means of a potentiometer or linear hall effect device, (which is more resistant to dust and grease) or it can also be automatically controlled for example by using a rotational detector such as a Gray code optical encoder.

Types of Drives

There are three general categories of electric drives: DC motor drives, eddy current drives and AC motor drives. Each of these general types can be further divided into numerous variations. Electric drives generally include both an electric motor and a speed control unit or system. The term *drive* is often applied to the controller without the motor. In the early days of electric drive technology, electromechanical control systems were used. Later, electronic controllers were designed using various types of vacuum tubes. As suitable solid state electronic components became available, new controller designs incorporated the latest electronic technology.

DC Drives

DC drives are DC motor speed control systems. Since the speed of a DC motor is directly proportional to armature voltage and inversely proportional to motor flux (which is a function of field current), either armature voltage or field current can be used to control speed.

Eddy Current Drives

An eddy current drive (sometimes called a Dynamatic drive, after one of the most common brand names) consists of a fixed speed motor (generally an induction motor) and an eddy current clutch. The clutch contains a fixed speed rotor and an adjustable speed rotor separated by a small air gap. A direct current in a field coil produces a magnetic field that determines the torque transmitted from the input rotor to the output rotor. The controller provides closed loop speed regulation by varying clutch current, only allowing the clutch to transmit enough torque to operate at the desired speed. Speed feedback is typically provided via an integral AC tachometer.

Eddy current drives are slip-controlled systems the slip energy of which is necessarily all dissipated as heat. Such drives are therefore generally less efficient than AC/DC-AC conversion based drives. The motor develops the torque required by the load and operates at full speed. The output shaft transmits the same torque to the load, but turns at a slower speed. Since power is proportional to torque multiplied by speed, the input power is proportional to motor speed times operating torque while the output power is output speed times operating torque. The difference between the motor speed and the output speed is called the *slip speed*. Power proportional to the slip speed times operating torque is dissipated as heat in the clutch.

While it has been surpassed by the variable-frequency drive in most variable-speed applications, the eddy current clutch is still often used to couple motors to high-inertia loads that are frequently stopped and started, such as stamping presses, conveyors, hoisting machinery, and some larger machine tools, allowing gradual starting, with less maintenance than a mechanical clutch or hydraulic transmission.

AC Drives

AC drives are AC motor speed control systems. A slip-controlled wound-rotor induction motor (WRIM) drive controls speed by varying motor slip via rotor slip rings either by electronically recovering slip power fed back to the stator bus or by varying the resistance of external resistors in the rotor circuit. Along with eddy current drives, resistance-based WRIM drives have lost popularity because they are less efficient than AC/DC-AC-based WRIM drives and are used only in special situations.

Slip energy recovery systems return energy to the WRIM's stator bus, converting slip energy and feeding it back to the stator supply. Such recovered energy would otherwise be wasted as heat in resistance-based WRIM drives. Slip energy recovery variable-speed drives are used in such applications as large pumps and fans, wind turbines, shipboard propulsion systems, large hydro-pumps/generators and utility energy storage flywheels. Early slip energy recovery systems using electromechanical components for AC/DC-AC conversion (i.e., consisting of rectifier, DC motor and AC generator) are termed Kramer drives, more recent systems using variable-frequency drives (VFDs) being referred to as static Kramer drives.

In general, a VFD in its most basic configuration controls the speed of an induction or synchronous motor by adjusting the frequency of the power supplied to the motor.

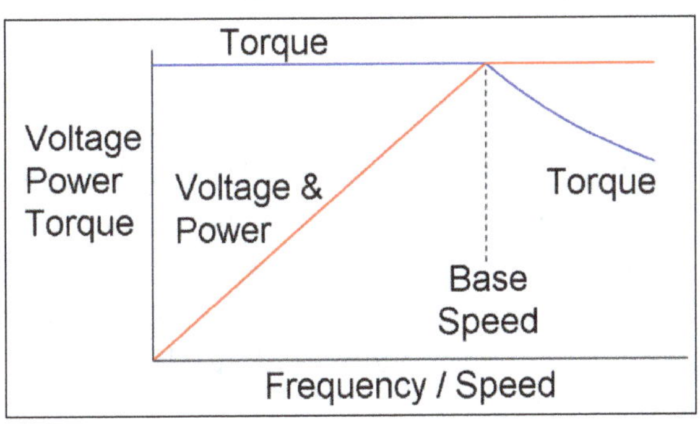

When changing VFD frequency in standard low-performance variable-torque applications using Volt-per-Hertz (V/Hz) control, the AC motor's voltage-to-frequency ratio can be maintained constant, and its power can be varied, between the minimum and maximum operating frequencies up to a base frequency. Constant voltage operation above base frequency, and therefore with reduced V/Hz ratio, provides reduced torque and constant power capability.

Regenerative AC drives are a type of AC drive which have the capacity to recover the braking energy of a load moving faster than the motor speed (an overhauling load) and return it to the power system.

Variable-frequency Drive

A variable-frequency drive (VFD) or adjustable-frequency drive (AFD), variable-voltage/variable-frequency (VVVF) drive, variable speed drive (VSD), AC drive, micro drive or inverter drive is a type of adjustable-speed drive used in electro-mechanical drive systems to control AC motor speed and torque by varying motor input frequency and voltage.

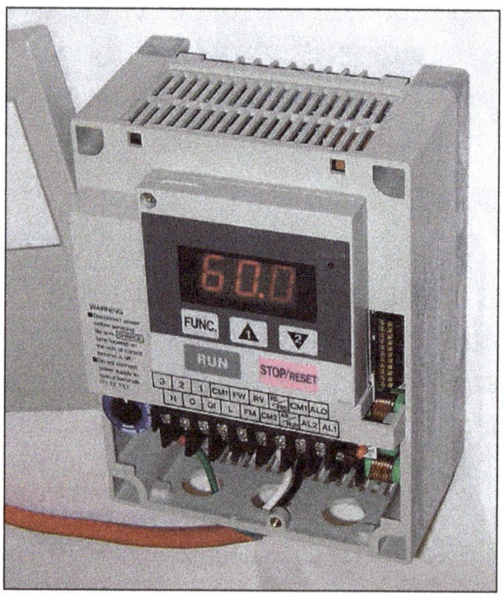

Small variable-frequency drive.

VFDs are used in applications ranging from small appliances to large compressors. About 25% of the world's electrical energy is consumed by electric motors in industrial applications. Systems using VFDs can be more efficient than those using throttling control of fluid flow, such as in systems with pumps or fans. However, the global market penetration for all applications of VFDs is relatively small. Over the last four decades, power electronics technology has reduced VFD cost and size and has improved performance through advances in semiconductor switching devices, drive topologies, simulation and control techniques, and control hardware and software. VFDs are made in a number of different low- and medium-voltage AC-AC and DC-AC topologies.

Chassis of above VFD (cover removed).

System Description and Operation

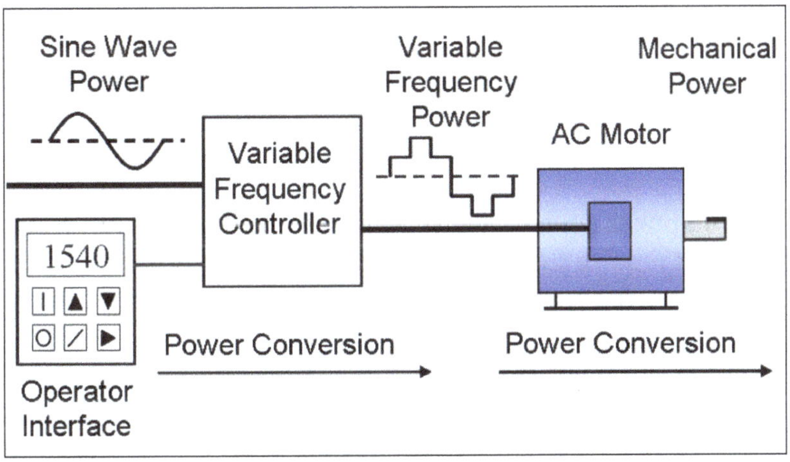

VFD system.

A variable-frequency drive is a device used in a drive system consisting of the following three main sub-systems: AC motor, main drive controller assembly, and drive/operator interface.

AC Motor

The AC electric motor used in a VFD system is usually a three-phase induction motor. Some types of single-phase motors or synchronous motors can be advantageous in some situations, but generally three-phase induction motors are preferred as the most economical. Motors that are designed for fixed-speed operation are often used. Elevated-voltage stresses imposed on induction motors that are supplied by VFDs require that such motors be designed for definite-purpose inverter-fed duty in accordance with such requirements as Part 31 of NEMA Standard MG-1.

Controller

The VFD controller is a solid-state power electronics conversion system consisting of three distinct sub-systems: a rectifier bridge converter, a direct current (DC) link, and an inverter. Voltage-source inverter (VSI) drives are by far the most common type of drives. Most drives are AC-AC drives in that they convert AC line input to AC inverter output. However, in some applications such as common DC bus or solar applications, drives are configured as DC-AC drives. The most basic rectifier converter for the VSI drive is configured as a three-phase, six-pulse, full-wave diode bridge. In a VSI drive, the DC link consists of a capacitor which smooths out the converter's DC output ripple and provides a stiff input to the inverter. This filtered DC voltage is converted to quasi-sinusoidal AC voltage output using the inverter's active switching elements. VSI drives provide higher power factor and lower harmonic distortion than phase-controlled current-source inverter (CSI) and load-commutated inverter (LCI) drives. The drive controller can also be configured as a phase converter having single-phase converter input and three-phase inverter output.

Controller advances have exploited dramatic increases in the voltage and current ratings and switching frequency of solid-state power devices over the past six decades. Introduced in 1983, the insulated-gate bipolar transistor (IGBT) has in the past two decades come to dominate VFDs as an inverter switching device.

In variable-torque applications suited for Volts-per-Hertz (V/Hz) drive control, AC motor characteristics require that the voltage magnitude of the inverter's output to the motor be adjusted to match the required load torque in a linear V/Hz relationship. For example, for 460 V, 60 Hz motors, this linear V/Hz relationship is 460/60 = 7.67 V/Hz. While suitable in wide-ranging applications, V/Hz control is sub-optimal in high-performance applications involving low speed or demanding, dynamic speed regulation, positioning, and reversing load requirements. Some V/Hz control drives can also operate in quadratic V/Hz mode or can even be programmed to suit special multi-point V/Hz paths.

The two other drive control platforms, vector control and direct torque control (DTC), adjust the motor voltage magnitude, angle from reference, and frequency so as to precisely control the motor's magnetic flux and mechanical torque.

Although space vector pulse-width modulation (SVPWM) is becoming increasingly popular, sinusoidal PWM (SPWM) is the most straightforward method used to vary drives' motor voltage (or current) and frequency. With SPWM control, quasi-sinusoidal, variable-pulse-width output is constructed from intersections of a saw-toothed carrier signal with a modulating sinusoidal signal which is variable in operating frequency as well as in voltage (or current).

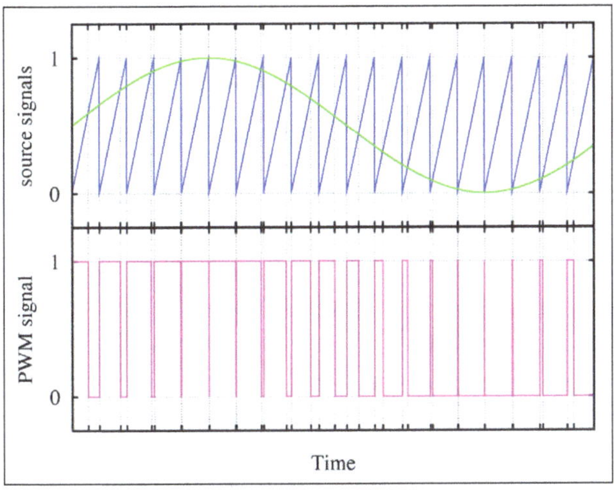

SPWM carrier-sine input & 2-level PWM output.

Operation of the motors above rated nameplate speed (base speed) is possible, but is limited to conditions that do not require more power than the nameplate rating of the motor. This is sometimes called "field weakening" and, for AC motors, means operating at less than rated V/Hz and above rated nameplate speed. Permanent magnet synchronous motors have quite limited field-weakening speed range due to the constant magnet flux linkage. Wound-rotor synchronous motors and induction motors have much wider speed range. For example, a 100 HP, 460 V, 60 Hz, 1775 RPM (4-pole) induction motor supplied with 460 V, 75 Hz (6.134 V/Hz), would be limited to 60/75 = 80% torque at 125% speed (2218.75 RPM) = 100% power. At higher speeds, the induction motor torque has to be limited further due to the lowering of the breakaway torque[a] of the motor. Thus, rated power can be typically produced only up to 130-150% of the rated nameplate speed. Wound-rotor synchronous motors can be run at even higher speeds. In rolling mill drives, often 200-300% of the base speed is used. The mechanical strength of the rotor limits the maximum speed of the motor.

An embedded microprocessor governs the overall operation of the VFD controller. Basic programming of the microprocessor is provided as user-inaccessible firmware. User programming of display, variable, and function block parameters is provided to control, protect, and monitor the VFD, motor, and driven equipment.

The basic drive controller can be configured to selectively include such optional power components and accessories as follows:

- Connected upstream of converter – circuit breaker or fuses, isolation contactor, EMC filter, line reactor, passive filter.
- Connected to DC link – braking chopper, braking resistor.
- Connected downstream of inverter—output reactor, sine wave filter, dV/dt filter.

Operator Interface

The operator interface provides a means for an operator to start and stop the motor and adjust the operating speed. The VFD may also be controlled by a programmable logic controller through Modbus or another similar interface. Additional operator control functions might include reversing, and switching between manual speed adjustment and automatic control from an external process control signal. The operator interface often includes an alphanumeric display or indication lights and meters to provide information about the operation of the drive. An operator interface keypad and display unit is often provided on the front of the VFD controller as shown in the photograph above. The keypad display can often be cable-connected and mounted a short distance from the VFD controller. Most are also provided with input and output (I/O) terminals for connecting push buttons, switches, and other operator interface devices or control signals. A serial communications port is also often available to allow the VFD to be configured, adjusted, monitored, and controlled using a computer.

Speed Control

There are two main ways to control the speed of a VFD; networked or hardwired. Networked involves transmitting the intended speed over a communication protocol such as Modbus, Modbus/TCP, EtherNet/IP, or via a keypad using Display Serial Interface while hardwired involves a pure electrical means of communication. Typical means of hardwired communication are: 4-20mA, 0-10VDC, or using the internal 24VDC power supply with a potentiometer. Speed can also be controlled remotely and locally. Remote control instructs the VFD to ignore speed commands from the keypad while local control instructs the VFD to ignore external control and only abide by the keypad. On some drives the same pins are used for both 0-10VDC and 4-20mA and are selected via a jumper.

Programming a VFD

Depending on the model a VFD's operating parameters can be programmed via: dedicated programming software, internal keypad, external keypad, or SD card. VFDs will often block out most programming changes while running. Typical parameters that need to be set include: motor nameplate information, speed reference source, on/off control source and braking control. It is also common for VFDs to provide debugging information such as fault codes and the states of the input signals.

Starting and Software Behavior

Most VFDs allow auto-starting to be enabled. Which will drive the output to a designated frequency after a power cycle, or after a fault has been cleared, or after the emergency stop signal has been restored (generally emergency stops are active low logic). One popular way to control a VFD is to enable auto-start and place L1, L2, and L3 into a contactor. Powering on the contactor thus turns on the drive and has it output to a designated speed. Depending on the sophistication of the drive multiple auto-starting behavior can be developed e.g. the drive auto-starts on power up but does not auto-start from clearing an emergency stop until a reset has been cycled.

Drive Operation

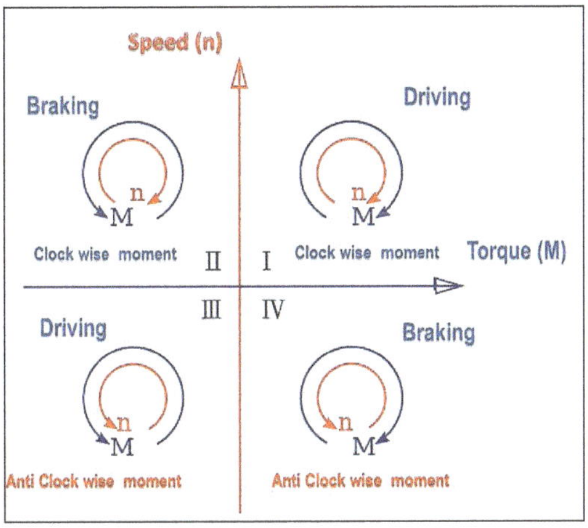

Electric motor speed-torque chart.

Referring to the accompanying chart, drive applications can be categorized as single-quadrant, two-quadrant, or four-quadrant; the chart's four quadrants are defined as follows:

- Quadrant I: Driving or motoring, forward accelerating quadrant with positive speed and torque.
- Quadrant II: Generating or braking, forward braking-decelerating quadrant with positive speed and negative torque.
- Quadrant III: Driving or motoring, reverse accelerating quadrant with negative speed and torque.
- Quadrant IV: Generating or braking, reverse braking-decelerating quadrant with negative speed and positive torque.

Most applications involve single-quadrant loads operating in quadrant I, such as in variable-torque (e.g. centrifugal pumps or fans) and certain constant-torque (e.g. extruders) loads.

Certain applications involve two-quadrant loads operating in quadrant I and II where the speed is positive but the torque changes polarity as in case of a fan decelerating faster than natural mechanical losses. Some sources define two-quadrant drives as loads operating in quadrants I and III where the speed and torque is same (positive or negative) polarity in both directions.

Certain high-performance applications involve four-quadrant loads (Quadrants I to IV) where the speed and torque can be in any direction such as in hoists, elevators, and hilly conveyors. Regeneration can occur only in the drive's DC link bus when inverter voltage is smaller in magnitude than the motor back-EMF and inverter voltage and back-EMF are the same polarity.

In starting a motor, a VFD initially applies a low frequency and voltage, thus avoiding high inrush current associated with direct-on-line starting. After the start of the VFD, the applied frequency and voltage are increased at a controlled rate or ramped up to accelerate the load. This starting method typically allows a motor to develop 150% of its rated torque while the VFD is drawing less than 50% of its rated current from the mains in the low-speed range. A VFD can be adjusted to produce a steady 150% starting torque from standstill right up to full speed. However, motor cooling deteriorates and can result in overheating as speed decreases such that prolonged low-speed operation with significant torque is not usually possible without separately motorized fan ventilation.

With a VFD, the stopping sequence is just the opposite as the starting sequence. The frequency and voltage applied to the motor are ramped down at a controlled rate. When the frequency approaches zero, the motor is shut off. A small amount of braking torque is available to help decelerate the load a little faster than it would stop if the motor were simply switched off and allowed to coast. Additional braking torque can be obtained by adding a braking circuit (resistor controlled by a transistor) to dissipate the braking energy. With a four-quadrant rectifier (active front-end), the VFD is able to brake the load by applying a reverse torque and injecting the energy back to the AC line.

Benefits

Energy Savings

Many fixed-speed motor load applications that are supplied direct from AC line power can save energy when they are operated at variable speed by means of VFD. Such energy cost savings are especially pronounced in variable-torque centrifugal fan and pump applications, where the load's torque and power vary with the square and cube, respectively, of the speed. This change gives a large power reduction compared to fixed-speed operation for a relatively small reduction in speed. For example, at 63% speed a motor load consumes only 25% of its full-speed power. This reduction is in accordance with affinity laws that define the relationship between various centrifugal load variables.

In the United States, an estimated 60-65% of electrical energy is used to supply motors, 75% of which are variable-torque fan, pump, and compressor loads. Eighteen percent of the energy used in the 40 million motors in the U.S. could be saved by efficient energy improvement technologies such as VFDs.

VVVF used on train of Guangzhou metro.

Only about 3% of the total installed base of AC motors are provided with AC drives. However, it is estimated that drive technology is adopted in as many as 30-40% of all newly installed motors.

An energy consumption breakdown of the global population of AC motor.

Table: Global population of motors.

	Small	General Purpose - Medium-Size	Large
Power	10 W - 750 W	0.75 kW - 375 kW	375 kW - 10000 kW
Phase, voltage	1-ph., <240 V	3-ph., 200 V to 1 kV	3-ph., 1 kV to 20 kV
% total motor energy	9%	68%	23%
Total stock	2 billion	230 million	0.6 million

Control Performance

AC drives are used to bring about process and quality improvements in industrial and commercial applications' acceleration, flow, monitoring, pressure, speed, temperature, tension, and torque.

Fixed-speed loads subject the motor to a high starting torque and to current surges that are up to eight times the full-load current. AC drives instead gradually ramp the motor up to operating speed to lessen mechanical and electrical stress, reducing maintenance and repair costs, and extending the life of the motor and the driven equipment.

Variable-speed drives can also run a motor in specialized patterns to further minimize mechanical and electrical stress. For example, an S-curve pattern can be applied to a conveyor application for smoother deceleration and acceleration control, which reduces the backlash that can occur when a conveyor is accelerating or decelerating.

Performance factors tending to favor the use of DC drives over AC drives include such requirements as continuous operation at low speed, four-quadrant operation with regeneration, frequent acceleration and deceleration routines, and need for the motor to be protected for a hazardous area. The following table compares AC and DC drives according to certain key parameters:

Drive type	DC	AC VFD	AC VFD	AC VFD	AC VFD
Control platform	Brush type DC	V/Hz control	Vector control	Vector control	Vector control
Control criteria	Closed-loop	Open-loop	Open-loop	Closed-loop	Open-loop w. HFI^
Motor	DC	IM	IM	IM	Interior PM
Typical speed regulation (%)	0.01	1	0.5	0.01	0.02
Typical speed range at constant torque (%)	0-100	10-100	3-100	0-100	0-100
Min. speed at 100% torque (% of base)	Standstill	8%	2%	Standstill	Standstill (200%)
Multiple-motor operation recommended	No	Yes	No	No	No
Fault protection (Fused only or inherent to drive)	Fused only	Inherent	Inherent	Inherent	Inherent
Maintenance	(Brushes)	Low	Low	Low	Low
Feedback device	Tachometer or encoder	N/A	N/A	Encoder	N/A

VFD Types and Ratings

Generic Topologies

AC drives can be classified according to the following generic topologies:

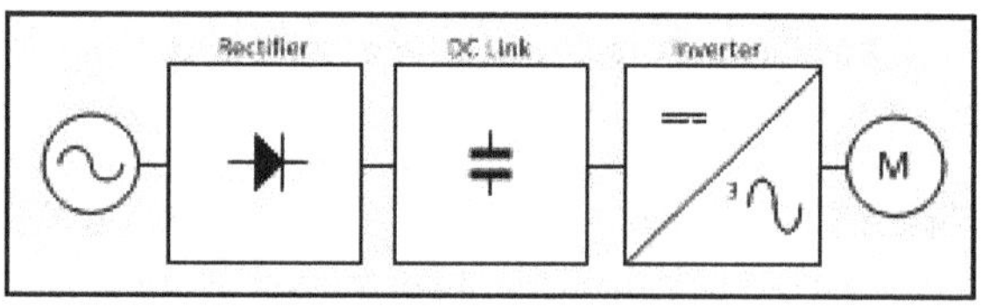

Topology of VSI drive.

- Voltage-source inverter (VSI) drive topologies: In a VSI drive, the DC output of the diode-bridge converter stores energy in the capacitor bus to supply stiff voltage input to the inverter. The vast majority of drives are VSI type with PWM voltage output.

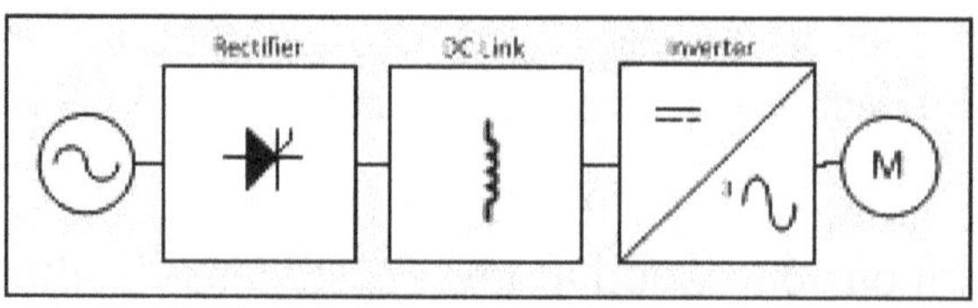

Topology of CSI drive.

- Current-source inverter (CSI) drive topologies: In a CSI drive, the DC output of the SCR-bridge converter stores energy in series-Inductor connection to supply stiff current input to the inverter. CSI drives can be operated with either PWM or six-step waveform output.

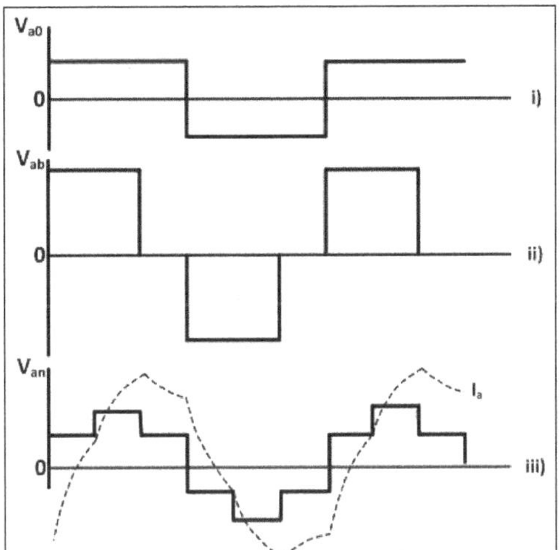

Six-step drive waveforms per VSI phase leg: i) across inverter bridge, ii) line-to-line, and iii) line-to-neutral

Six-step drive waveforms.

- Six-step inverter drive topologies: Now largely obsolete, six-step drives can be either VSI or CSI type and are also referred to as variable-voltage inverter drives, pulse-amplitude modulation (PAM) drives, square-wave drives or D.C. chopper inverter drives. In a six-step drive, the DC output of the SCR-bridge converter is smoothed via capacitor bus and series-reactor conncction to supply via Darlington Pair or IGBT inverter quasi-sinusoidal, six-step voltage or current input to an induction motor.
- Load commutated inverter (LCI) drive topologies: In an LCI drive (a special CSI case), the DC output of the SCR-bridge converter stores energy via DC link inductor circuit to supply stiff quasi-sinusoidal six-step current output of a second SCR-bridge's inverter and an over-excited synchronous machine.

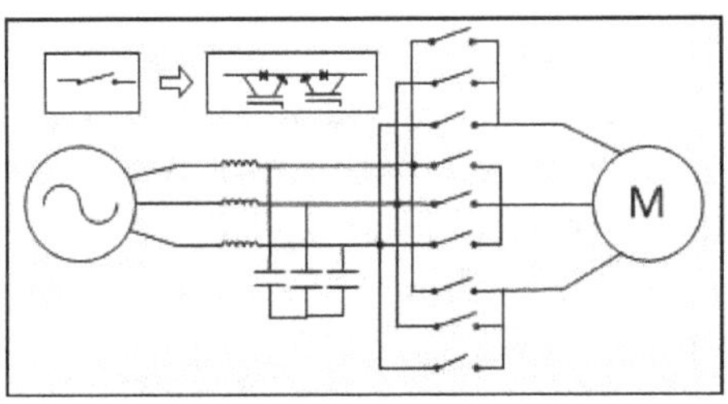

Topology of direct matrix converter.

- Cycloconverter or matrix converter (MC) topologies: Cycloconverters and MCs are AC-AC converters that have no intermediate DC link for energy storage. A cycloconverter operates as a three-phase current source via three anti-parallel-connected SCR-bridges in six-pulse configuration, each cycloconverter phase acting selectively to convert fixed line frequency AC voltage to an alternating voltage at a variable load frequency. MC drives are IGBT-based.

- Doubly fed slip recovery system topologies: A doubly fed slip recovery system feeds rectified slip power to a smoothing reactor to supply power to the AC supply network via an inverter, the speed of the motor being controlled by adjusting the DC current.

Control Platforms

Most drives use one or more of the following control platforms:

- PWM V/Hz scalar control.
- PWM field-oriented control (FOC) or vector control.
- Direct torque control (DTC).

Load Torque and Power Characteristics

Variable-frequency drives are also categorized by the following load torque and power characteristics:

- Variable torque, such as in centrifugal fan, pump, and blower applications.
- Constant torque, such as in conveyor and positive-displacement pump applications.
- Constant power, such as in machine tool and traction applications.

Available Power Ratings

VFDs are available with voltage and current ratings covering a wide range of single-phase and multi-phase AC motors. Low-voltage (LV) drives are designed to operate at output voltages equal to or less than 690 V. While motor-application LV drives are available in ratings of up to the order of 5 or 6 MW, economic considerations typically favor medium-voltage (MV) drives with much lower power ratings. Different MV drive topologies are configured in accordance with the voltage/current-combination ratings used in different drive controllers' switching devices such that any given voltage rating is greater than or equal to one to the following standard nominal motor voltage ratings: generally either 2 $^{3}/_{4}$.16 kV (60 Hz) or 3 $^{3}/_{6}$.6 kV (50 Hz), with one thyristor manufacturer rated for up to 12 kV switching. In some applications a step-up transformer is placed between a LV drive and a MV motor load. MV drives are typically rated for motor applications greater than between about 375 and 750 kW (503 and 1,006 hp). MV drives have historically required considerably more application design effort than required for LV drive applications. The power rating of MV drives can reach 100 MW (130,000 hp), a range of different drive topologies being involved for different rating, performance, power quality, and reliability requirements.

Application Considerations

AC Line Harmonics

While harmonics in the PWM output can easily be filtered by carrier-frequency-related filter inductance to supply near-sinusoidal currents to the motor load, the VFD's diode-bridge rectifier converts AC line voltage to DC voltage output by super-imposing non-linear half-phase current pulses thus creating harmonic current distortion, and hence voltage distortion, of the AC line input. When the VFD loads are relatively small in comparison to the large, stiff power system available from the electric power company, the effects of VFD harmonic distortion of the AC grid can often be within acceptable limits. Furthermore, in low-voltage networks, harmonics caused by single-phase equipment such as computers and TVs are partially cancelled by three-phase diode bridge harmonics because their 5th and 7th harmonics are in counterphase. However, when the proportion of VFD and other non-linear load compared to total load or of non-linear load compared to the stiffness at the AC power supply, or both, is relatively large enough, the load can have a negative impact on the AC power waveform available to other power company customers in the same grid.

When the power company's voltage becomes distorted due to harmonics, losses in other loads such as normal fixed-speed AC motors are increased. This condition may lead to overheating and shorter operating life. Also, substation transformers and compensation capacitors are affected negatively. In particular, capacitors can cause resonance conditions that can unacceptably magnify harmonic levels. In order to limit the voltage distortion, owners of VFD load may be required to install filtering equipment to reduce harmonic distortion below acceptable limits.

Alternatively, the utility may adopt a solution by installing filtering equipment of its own at substations affected by the large amount of VFD equipment being used. In high-power installations, harmonic distortion can be reduced by supplying multi-pulse rectifier-bridge VFDs from transformers with multiple phase-shifted windings.

It is also possible to replace the standard diode-bridge rectifier with a bi-directional IGBT switching device bridge mirroring the standard inverter which uses IGBT switching device output to the motor. Such rectifiers are referred to by various designations including active infeed converter (AIC), active rectifier, IGBT supply unit (ISU), active front end (AFE), or four-quadrant operation. With PWM control and a suitable input reactor, an AFE's AC line current waveform can be nearly sinusoidal. AFE inherently regenerates energy in four-quadrant mode from the DC side to the AC grid. Thus, no braking resistor is needed, and the efficiency of the drive is improved if the drive is frequently required to brake the motor.

Two other harmonics mitigation techniques exploit use of passive or active filters connected to a common bus with at least one VFD branch load on the bus. Passive filters involve the design of one or more low-pass LC filter traps, each trap being tuned as required to a harmonic frequency (5th, 7th, 11th, 13th,... kq+/-1, where k=integer, q=-pulse number of converter).

It is very common practice for power companies or their customers to impose harmonic distortion limits based on IEC or IEEE standards. For example, IEEE Standard 519 limits at the customer's connection point call for the maximum individual frequency voltage harmonic to be no more than 3% of the fundamental and call for the voltage total harmonic distortion (THD) to be no more than 5% for a general AC power supply system.

Switching Frequency Foldback

One drive uses a default switching frequency setting of 4 kHz. Reducing the drive's switching frequency (the carrier-frequency) reduces the heat generated by the IGBTs.

A carrier frequency of at least ten times the desired output frequency is used to establish the PWM switching intervals. A carrier frequency in the range of 2,000 to 16,000 Hz is common for LV [low voltage, under 600 Volts AC] VFDs. A higher carrier frequency produces a better sine wave approximation but incurs higher switching losses in the IGBT, decreasing the overall power conversion efficiency.

Noise Smoothing

Some drives have a noise smoothing feature that can be turned on to introduce a random variation to the switching frequency. This distributes the acoustic noise over a range of frequencies to lower the peak noise intensity.

Long-lead Effects

The carrier-frequency pulsed output voltage of a PWM VFD causes rapid rise times in these pulses, the transmission line effects of which must be considered. Since the transmission-line impedance of the cable and motor are different, pulses tend to reflect back from the motor terminals into the cable. The resulting reflections can produce overvoltages equal to twice the DC bus voltage or up to 3.1 times the rated line voltage for long cable runs, putting high stress on the cable and motor windings, and eventual insulation failure. Insulation standards for three-phase motors rated 230 V or less adequately protect against such long-lead overvoltages. On 460 V or 575 V systems and inverters with 3rd-generation 0.1-microsecond-rise-time IGBTs, the maximum recommended cable distance between VFD and motor is about 50 m or 150 feet. For emerging SiC MOSFET powered drives, significant overvoltages have been observed at cable lengths as short as 3 meters. Solutions to overvoltages caused by long lead lengths include minimizing cable length, lowering carrier frequency, installing dV/dt filters, using inverter-duty-rated motors (that are rated 600 V to withstand pulse trains with rise time less than or equal to 0.1 microsecond, of 1,600 V peak magnitude), and installing LCR low-pass sine wave filters. Selection of optimum PWM carrier frequency for AC drives involves balancing noise, heat, motor insulation stress, common-mode voltage-induced motor bearing current damage, smooth motor operation, and other factors. Further harmonics attenuation can be obtained by using an LCR low-pass sine wave filter or dV/dt filter.

Motor Bearing Currents

Carrier frequencies above 5 kHz are likely to cause bearing damage unless protective measures are taken.

PWM drives are inherently associated with high-frequency common-mode voltages and currents which may cause trouble with motor bearings. When these high-frequency voltages find a path to earth through a bearing, transfer of metal or electrical discharge machining (EDM) sparking occurs between the bearing's ball and the bearing's race. Over time, EDM-based sparking causes erosion in the bearing race that can be seen as a fluting pattern. In large motors, the stray capacitance of the windings provides paths for high-frequency currents that pass through the motor shaft ends, leading to a circulating type of bearing current. Poor grounding of motor stators can lead to shaft-to-ground bearing currents. Small motors with poorly grounded driven equipment are susceptible to high-frequency bearing currents.

Prevention of high-frequency bearing current damage uses three approaches: good cabling and grounding practices, interruption of bearing currents, and filtering or damping of common-mode currents for example through soft magnetic cores, the so-called inductive absorbers. Good cabling and grounding practices can include use of shielded, symmetrical-geometry power cable to supply the motor, installation of shaft grounding brushes, and conductive bearing grease. Bearing currents can be interrupted by installation of insulated bearings and specially designed electrostatic-shielded induction motors. Filtering and damping high-frequency bearing can be done though inserting soft magnetic cores over the three phases giving a high frequency impedance against the common mode or motor bearing currents. Another approach is to use instead of standard 2-level inverter drives, using either 3-level inverter drives or matrix converters.

Since inverter-fed motor cables' high-frequency current spikes can interfere with other cabling in facilities, such inverter-fed motor cables should not only be of shielded, symmetrical-geometry design but should also be routed at least 50 cm away from signal cables.

Dynamic Braking

Torque generated by the drive causes the induction motor to run at synchronous speed less the slip. If the load drives the motor faster than synchronous speed, the motor acts as a generator, converting mechanical power back to electrical power. This power is returned to the drive's DC link element (capacitor or reactor). A DC-link-connected electronic power switch or braking DC chopper controls dissipation of this power as heat in a set of resistors. Cooling fans may be used to prevent resistor overheating.

Dynamic braking wastes braking energy by transforming it to heat. By contrast, regenerative drives recover braking energy by injecting this energy into the AC line. The capital cost of regenerative drives is, however, relatively high.

Regenerative Drives

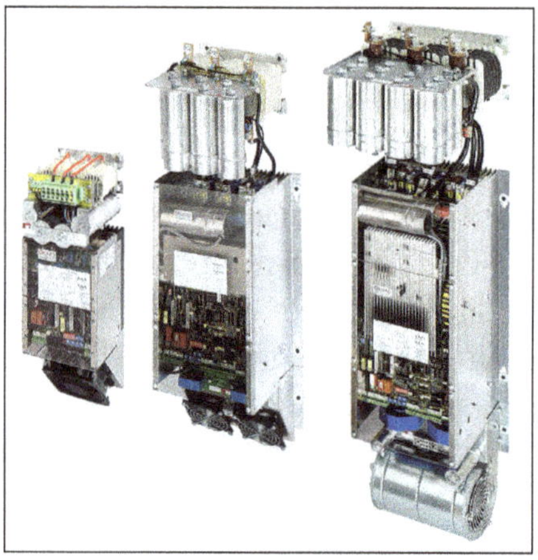

Line regenerative variable frequency drives, showing capacitors (top cylinders) and inductors attached, which filter the regenerated power.

Regenerative AC drives have the capacity to recover the braking energy of a load moving faster than the designated motor speed (an overhauling load) and return it to the power system.

Cycloconverter, Scherbius, matrix, CSI, and LCI drives inherently allow return of energy from the load to the line, while voltage-source inverters require an additional converter to return energy to the supply. Regeneration is useful in VFDs only where the value of the recovered energy is large compared to the extra cost of a regenerative system, and if the system requires frequent braking and starting. Regenerative VFDs are widely used where speed control of overhauling loads is required.

Some examples:

- Conveyor belt drives for manufacturing, which stop every few minutes. While stopped, parts are assembled correctly; once that is done, the belt moves on.
- A crane, where the hoist motor stops and reverses frequently, and braking is required to slow the load during lowering.
- Plug-in and hybrid electric vehicles of all types.

DC Drive

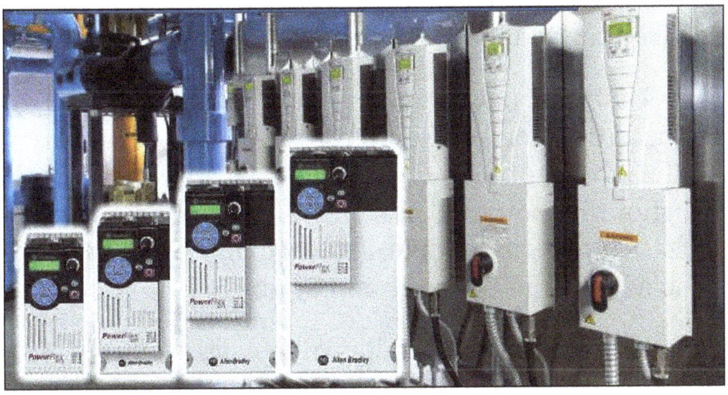

DC drive is basically a DC motor speed control system that supplies the voltage to the motor to operate at desired speed. Earlier, the variable DC voltage for the speed control of an industrial DC motor was generated by a DC generator.

By using an induction motor, the DC generator was driven at a fixed speed and by varying the field of the generator, variable voltage was generated. Soon after this Ward Leonard set was replaced by a mercury arc rectifier and later by thyristor converters. Nowadays, the thyristor family of devices is used widely to control the speed of the DC motor.

Components of a DC Drive

The main components of a DC drive system are shown in figure below:

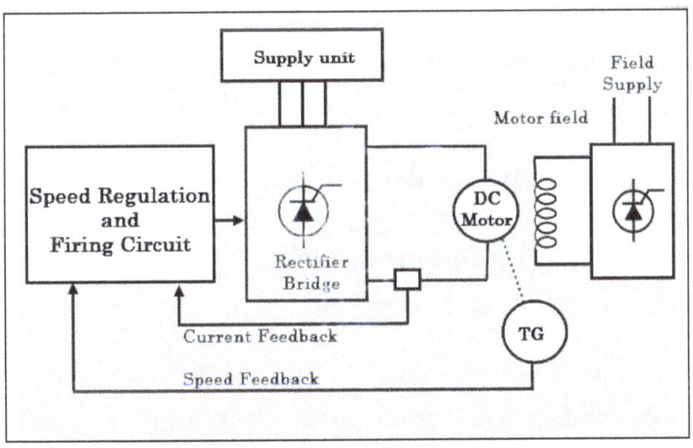

- DC Drive Input: Some thyristor based DC drives operate on a single phase supply and use four thyristors for full wave rectification. For larger motors, three phase power supply is needed because the waveforms are much smoother. In such cases, six thyristors are needed for full wave rectification.
- Rectifier Bridge: The power component of a controlled DC drive is a full wave bridge rectifier which can be driven by three phase or single phase supply. As mentioned above the number of thyristor may vary depends on the supply voltage.

- A six-thyristor bridge (in case of three phase converter) rectifies the incoming AC supply to DC supply to the motor armature. The firing angle control of these thyristors varies the voltage to the motor.

- Field Supply Unit: The power to be applied to the field winding is much lower than the armature power, so, most often single phase supply is provided. A separate thyristor bridge or diode rectifier is used for supplying the power to the field winding of the motor. In many cases a two-phase supply is drawn from the three phase input (that supplies power to the armature) and hence the field exciter is included in the armature supply unit.

  The function of the field supply unit is to provide a constant voltage to the field winding to create a constant field or flux in the motor. In some cases, this unit is supplied with thyristors to reduce the voltage applied to the field so as to control the speed of the motor above the base speed.

  In case of permanent magnet DC motors, the field supply unit is not included in the drive.

- Speed Regulation unit: It compares the operator instruction (desired speed) with feedback signals and sends appropriate signals to the firing circuit. In analog drives, this regulator unit consists of both voltage and current regulators. The voltage regulator accepts the speed error as input and produces the voltage output which is then applied to the current regulator.

  The current regulator then produces required firing current to the firing circuit. If more speed is required, additional current is called from the voltage regulator and hence thyristors conducts for more periods. Generally, this regulation (both voltage and current) is accomplished with proportional-integral-derivative controllers.

  The field current regulator is also provided where speed greater than the base speed is required. In modern digital microprocessor based drives, the speed control is achieved with a lookup table to determine the current for the firing circuit with additional digital circuitry.

- Firing Circuit: It supplies the gate pulses to thyristors so that they turned ON for particular periods to produce variable armature voltage. Isolation is also provided in this gate drive circuit.

Working Principle of DC Drives

In DC motors, the speed is proportional to the armature voltage and inversely proportional to the field current. And also, the armature current is proportional to the motor torque. Therefore, by increasing or reducing the applied voltage, the speed of the motor is varied. However, it is possible up to the rated voltage. If the speed greater than the base speed is required, the field current of the motor has to be reduced.

By reducing the field current, the flux in the motor reduces. The reduction of field current reduces the armature counter emf. The more armature current flows if there is less counter armature emf. Further, this armature current increases the motor torque and hence the speed. These are the two basic principles employed in DC drives to control the speed of the motor.

In armature controlled DC drives, drive unit provides a rated current and torque at any speed between zero and the base of the motor. By varying the armature voltage, variable speed is obtained as shown in figure.

Generally, a fixed field supply is provided in these DC drives. As the torque is constant (which describes a load type) over the speed range, the motor output horsepower is proportional to the speed (HP = T × N /525). The motor characteristics of this drive are shown below:

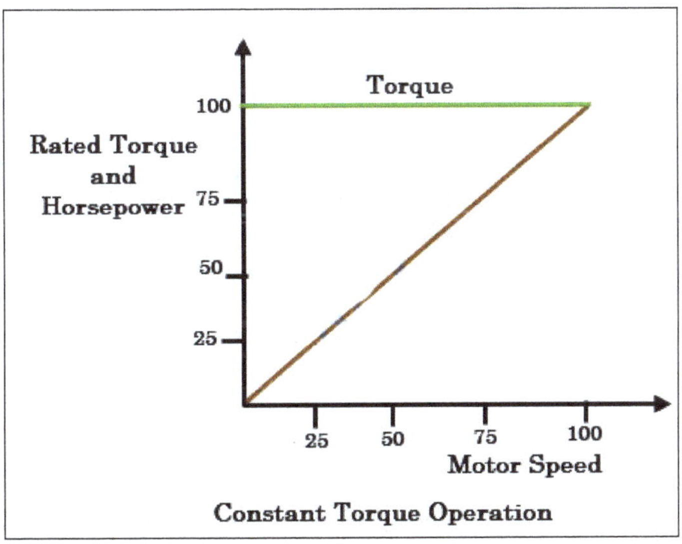

Constant Torque Operation.

In case of armature and field controlled drives, the armature voltage to the motor is controlled for constant torque-variable HP operation up to the base speed of the motor. And for the above base speed operation, drive switches to the field control for constant HP-reduced torque operation up to maximum speed as shown in figure below. In this case, reducing the field current increases the speed of the motor up to its maximum speed.

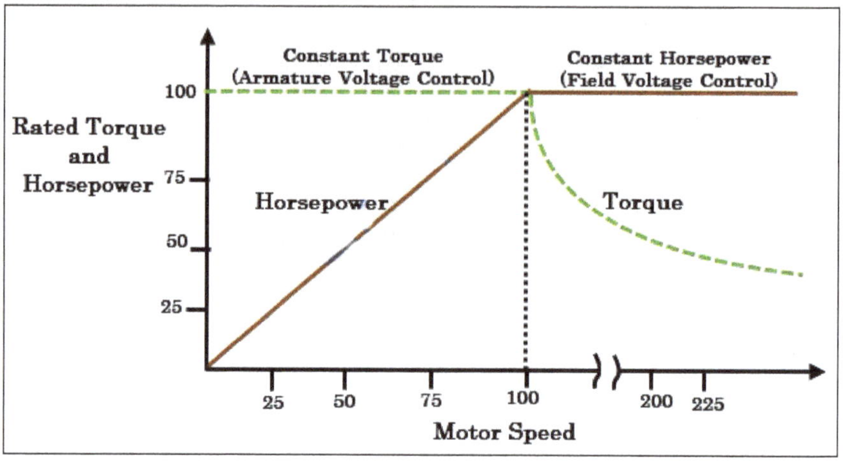

Motor Speed.

Digital and Analog DC Drives

Nowadays, digital implementations have replaced analog circuitry of electric drive system in all forms of industrial control. Digital controllers offer greater flexibility to produce the precise control, self-tuning, and ease of interfacing with host computers and other drives. However, a basic understanding of analogue version DC drive makes less difficult to understand its digital equivalent. Let us look on both of these DC drives.

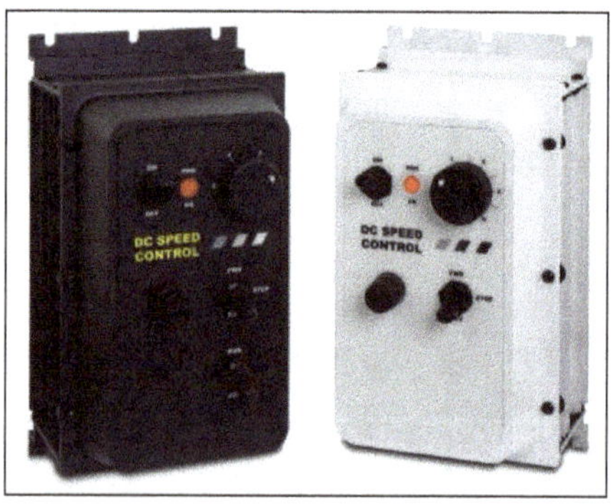

Analog DC Drives

A standard analog DC drive with speed and current control is shown in figure. The objective of this system is to provide speed control and hence the speed reference becomes the input to the system and speed of the motor is the output of the system which is measured by the tachometer.

The working of this drive goes like this; consider that motor is running at a set speed. Now, the speed reference signal has increased to somewhat greater than the actual speed. So there will be an error speed signal at left-hand summing junction as shown in figure. This speed error indicates the required acceleration by the motor, which means the torque and hence more current.

The error is amplified by the speed controller (which is basically a speed-error amplifier) and its output is given as current input reference to the inner control system. As the current reference increases, the inner current controller drives the more current to the motor thereby extra torque is provided.

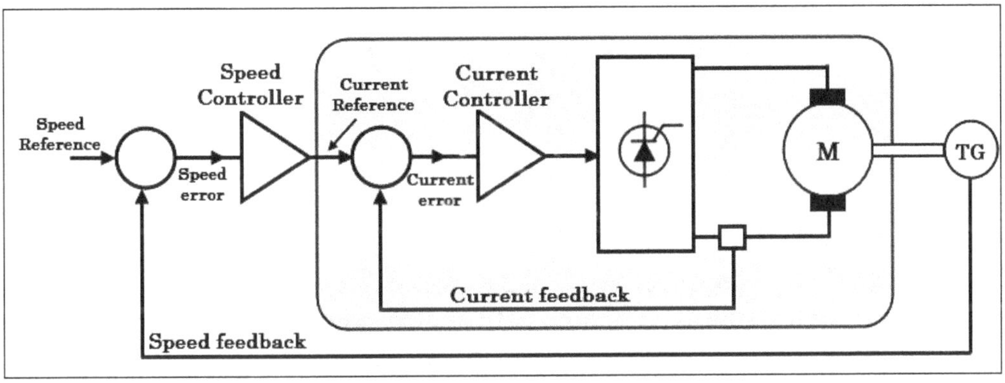

Analog DC Drives Block Diagram.

The inner current loop is responsible for maintaining the zero current error between the actual motor current and current reference signal which means to make actual motor current to follow the reference current. The amplified current error signal from the current controller controls the firing angle of the bridge and hence the output voltage of the converter. The current feedback is achieved either by DC transformer or by AC transformer (with rectifier) in the main supply lines.

This entire operation is performed by a current error amplifier with a high gain. In most cases, this amplifier is of proportional plus integral control (PI) type circuit that maintains the actual and desired currents exactly equal under steady-state conditions. This current controller also limits the current through the motor by considering the minimum and maximum currents of the motor.

The outer loop provides the speed control by comparing the actual speed obtained by the DC tachogenerator with desired or required speed from the speed reference. These two inputs are fed into the speed-error amplifier, and then resulted error is amplified and applied as an input to the current controller.

The speed amplifier produces the current output proportion to the speed error. For this amplifier also a PI control is employed (by using analog electronics) in order to achieve zero steady state error. Using this, the actual speed of the motor is maintained exactly at reference speed for all loads.

Digital DC Drives

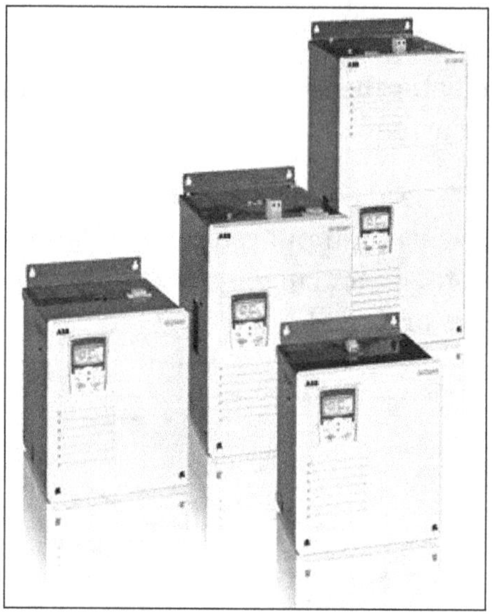

With the advancements in digital control, DC drives become more flexible and faster (due to faster response times) compared with analog drives. A schematic arrangement of digital DC drive is shown in below figure; of course it is similar to the analog scheme, but here analog circuit (analog amplifiers) is replaced by digital circuitry.

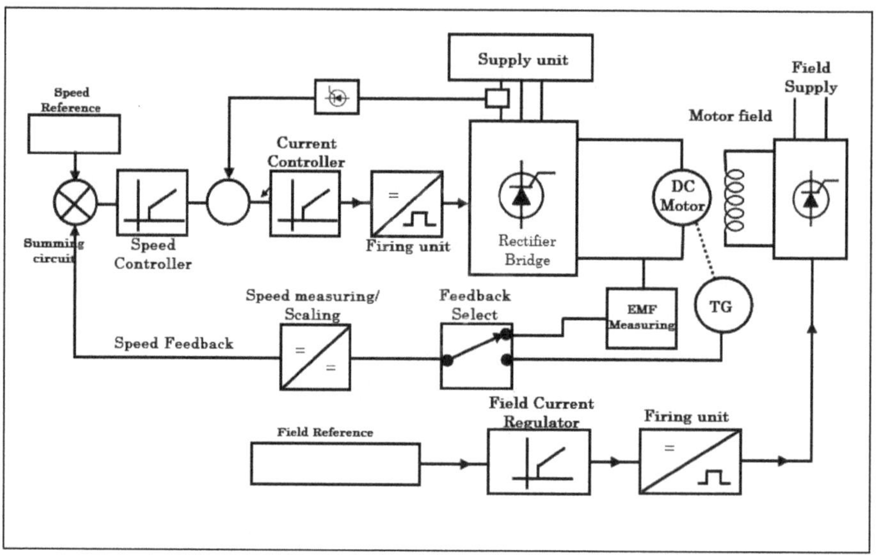

Digital DC Drive Block Diagram.

A speed reference signal given as the drive's input compared with the feedback speed in the summing circuit. If the output of the summing circuit is positive error, indicating that a speed increase is required and if it generates a negative error, indicating that a speed decrease is required (because motor is operating at faster than desired speed).

The error speed is given to the speed controller in the microprocessor which determines output voltage to operate the motor at desired speed. At the same time, current controller in the microprocessor determines the firing signals to the SCRs in the bridge converter. SCRs then convert the three phase supply to DC supply in relation to the desired speed.

This drive can operate in open loop without any feedback and can achieve a speed regulation of 5-8%. However, a speed regulation less than 5% is required in many applications. In such cases, the speed measuring/scaling unit switches to the EMF feedback measuring circuit. This feedback circuit measures the armature voltage, scales it in proportion to the output voltage (scaling function in microprocessor) and gives to the summing circuit. Further, it is transformed into a speed error signal in speed controller.

If the speed regulation less than 1% is required, tachometer generator feedback is used. So the speed measuring/scaling circuit then switches to the tachometer feedback. This feedback achieves very precise control compared with EMF feedback.

Also for field control (above rated speeds), this drive includes a separate field exciter. A field current regulator in the microprocessor determines the voltage to the field windings by accepting the flux/field reference signal from the operator. This regulator provides the firing signals required by the field converter unit to produce the required DC voltage proportional to the speed.

Silicon Controlled Rectifiers (SCRs) are widely used thyristors for large DC motor drives in its power conversion unit. An SCR conducts when a small voltage applied to its gate terminal. Its conduction continues till the starting of negative cycle and it turned OFF automatically once the voltage across the SCR goes through natural zero till next gated signal. The purpose of using these SCRs in DC drives is to convert the fixed AC supply to variable DC supply that controls the motor speed.

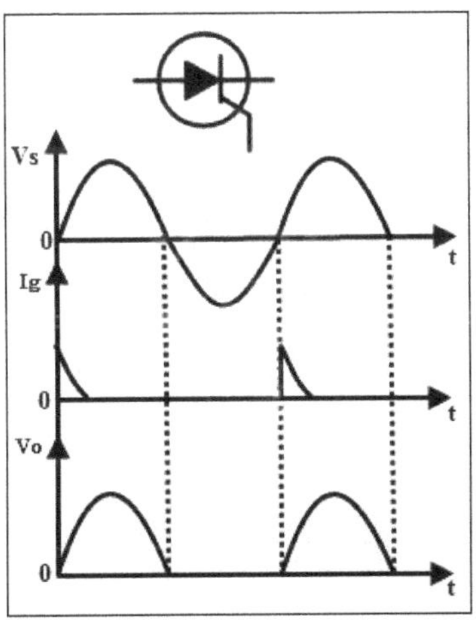

Some SCR DC drives are supplied from single phase AC supply and use four SCRs in the form of bridge for the DC rectification. In case of high power DC drives, a three phase supply with six SCRs is used for DC rectification. In case four quadrant operation (forward motoring, forward braking, reverse motoring and reverse braking) of the DC drive, a bridge rectifier of consisting of 12 SCRs with a three phase incoming supply is used. During each quadrant operation, SCRs are triggered at a phase angle in order to provide required DC voltage to the motor.

The connection of SCRs (for four quadrant operation of the drive) from incoming three phase AC supply to the DC output is shown in figure below. In this, the motoring SCR bridge and regeneration SCR bridge achieve the drive four quadrant operation by receiving the appropriate gate signals from (analog or digital) controller.

If the SCRs were gated with a phase angle of zero degrees, then the drive function as a rectifier which feds the full rectified rated DC supply to the motor and by varying the firing angle to the SCRs, a variable DC supply is applied to the motor.

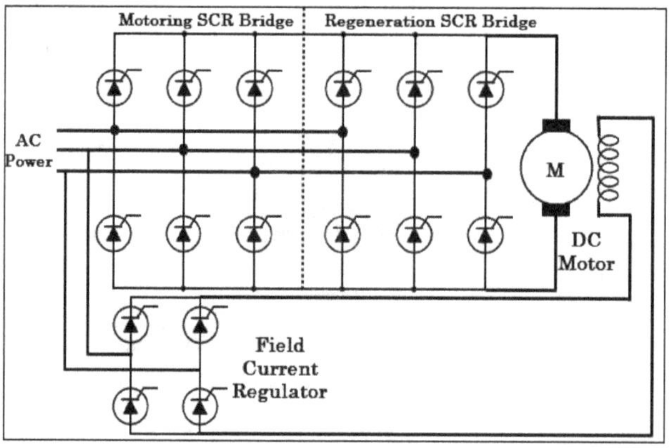

The DC output voltage waveform in relation to the AC waveform for above circuit is shown below. This average DC output voltage is obtained for 40°, 32° and 24° firing phase angles. By this way, the average output is controlled by varying the firing phase angles to the SCRs.

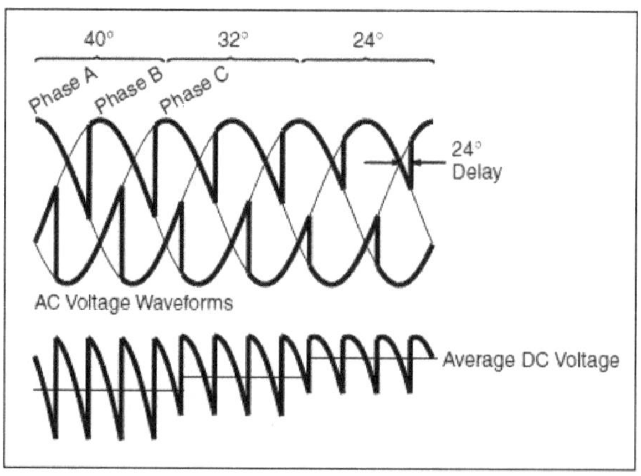

As the field winding also requires the regulated DC supply, only four SCRs are used in the field bridge converter. This is because field never requires a negative current and hence another set of SCRs is not required, which were used in armature for reversing the motor.

In modern DC drives, SCRs are completely replaced by MOSFETs and IGBTs in order to achieve high speed switching so that distortion to the AC incoming power and currents during switching is eliminated. Hence, the drive becomes more efficient and accurate. DC drives are also available in modular units that consisting analog and digital I/Os, multifunctional keypads, remote operator panels in addition to the software programming and configuring capabilities, from various manufacturers such as ABB, Siemens, Rockwell automation, Emerson, etc. These can be connected to the other drives or a computer host via communication cables.

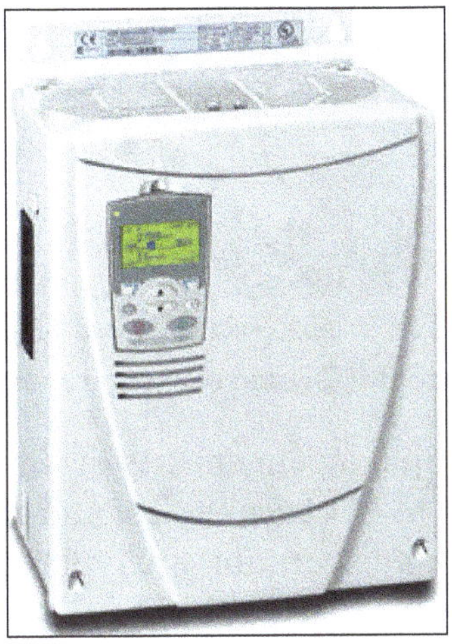

Programming macros of these drives enables to implement any control structure to an application. These are also capable of receiving the remote control signals from remote programmable logic controllers via field bus communication systems.

Programmable Logic Controller

A programmable logic controller (PLC) or programmable controller is an industrial digital computer which has been ruggedized and adapted for the control of manufacturing processes, such as assembly lines, or robotic devices, or any activity that requires high reliability control and ease of programming and process fault diagnosis.

PLCs were first developed in the automobile manufacturing industry to provide flexible, ruggedized and easily programmable controllers to replace hard-wired relays, timers and sequencers. Since then, they have been widely adopted as high-reliability automation controllers suitable for harsh environments. A PLC is an example of a "hard" real-time system since output results must be produced in response to input conditions within a limited time, otherwise unintended operation will result.

PLCs can range from small modular devices with tens of inputs and outputs (I/O), in a housing integral with the processor, to large rack-mounted modular devices with a count of thousands of I/O, and which are often networked to other PLC and SCADA systems.

PLC system in a rack, left-to-right: power supply unit (PSU), CPU, interface module (IM) and communication processor (CP).

They can be designed for multiple arrangements of digital and analog I/O, extended temperature ranges, immunity to electrical noise, and resistance to vibration and impact. Programs to control machine operation are typically stored in battery-backed-up or non-volatile memory.

It was from the automotive industry in the USA that the PLC was born. Before the PLC, control, sequencing, and safety interlock logic for manufacturing automobiles was mainly composed of relays, cam timers, drum sequencers, and dedicated closed-loop controllers. Since these could number in the hundreds or even thousands, the process for updating such facilities for the yearly model change-over was very time consuming and expensive, as electricians needed to individually rewire the relays to change their operational characteristics.

When digital computers became available, being general-purpose programmable devices, they were soon applied to control sequential and combinatorial logic in industrial processes. However these early computers required specialist programmers and stringent operating environmental control for temperature, cleanliness, and power quality. To meet these challenges the PLC was developed with several key attributes. It would tolerate the shop-floor environment, it would support discrete (bit-form) input and output in an easily extensible manner, it would not require years of training to use, and it would permit its operation to be monitored. Since many industrial processes have timescales easily addressed by millisecond response times, modern (fast, small, reliable) electronics greatly facilitate building reliable controllers, and performance could be traded off for reliability.

In 1968 GM Hydramatic (the automatic transmission division of General Motors) issued a request for proposals for an electronic replacement for hard-wired relay systems based on a white paper written by engineer Edward R. Clark. The winning proposal came from Bedford Associates of Bedford, Massachusetts. The first PLC, designated the 084 because it was Bedford Associates' eighty-fourth project, was the result. Bedford Associates started a new company dedicated to developing, manufacturing, selling, and servicing this new product: Modicon, which stood for modular digital controller. One of the people who worked on that project was Dick Morley, who is considered to be the "father" of the PLC. The Modicon brand was sold in 1977 to Gould Electronics, later acquired by German Company AEG, and then by French Schneider Electric, the current owner.

One of the very first 084 models built is now on display at Schneider Electric's facility in North Andover, Massachusetts. It was presented to Modicon by GM, when the unit was retired after nearly twenty years of uninterrupted service. Modicon used the 84 moniker at the end of its product range until the 984 made its appearance.

The automotive industry is still one of the largest users of PLCs.

In a parallel development Odo Josef Struger is sometimes known as the "father of the programmable logic controller" as well. He was involved in the invention of the Allen-Bradley programmable logic controller (PLC) during 1958 to 1960. Struger is credited with creating the PLC acronym. Allen-Bradley (now a brand owned by Rockwell Automation), the manufacturer of the controller, became a major programmable logic controller device manufacturer in the United States during the tenure of Struger.

Early PLCs were designed to replace relay logic systems. Relay systems were large and bulky and had an assortment of issues. The hard-wired nature made it difficult for design engineers to alter the process. Even small changes would require rewiring and careful updating of the documentation. If even one wire were out of place, or one relay failed, the whole system would become faulty. Often times technicians would spend hours troubleshooting by examining the schematics and comparing them to existing wiring. It was for this reason PLCs were programmed in "ladder logic", which strongly resembles a schematic diagram of relay logic. This program notation was chosen to reduce training demands for the existing technicians. Other early PLCs used a form of instruction list programming, based on a stack-based logic solver.

Modern PLCs can be programmed in a variety of ways, from the relay-derived ladder logic to programming languages such as specially adapted dialects of BASIC and C. Another method is state logic, a very high-level programming language designed to program PLCs based on state transition diagrams. The majority of PLC systems today adhere to the IEC 61131/3 control systems programming standard that defines 5 languages: Ladder Diagram (LD), Structured Text (ST), Function Block Diagram (FBD), Instruction List (IL) and sequential function chart (SFC).

Many early PLCs did not have accompanying programming terminals that were capable of graphical representation of the logic, and so the logic was instead represented as a series of logic expressions in some version of Boolean format, similar to Boolean algebra. As programming terminals evolved, it became more common for ladder logic to be used, for the aforementioned reasons and because it was a familiar format used for electro-mechanical control panels. Newer formats such as state logic and Function Block (which is similar to the way logic is depicted when using digital integrated logic circuits) exist, but they are still not as popular as ladder logic. A primary reason for this is that PLCs solve the logic in a predictable and repeating sequence, and ladder logic allows the programmer (the person writing the logic) to see any issues with the timing of the logic sequence more easily than would be possible in other formats.

Programming

PLC programs are typically written in a special application on a personal computer, then downloaded by a direct-connection cable or over a network to the PLC. The program is stored in the PLC either in battery-backed-up RAM or some other non-volatile flash memory. Often, a single PLC can be programmed to replace thousands of relays.

Early PLCs, up to the mid-1990s, were programmed using proprietary programming panels or special-purpose programming terminals, which often had dedicated function keys representing the various logical elements of PLC programs. Some proprietary programming terminals displayed the elements of PLC programs as graphic symbols, but plain ASCII character representations of contacts, coils, and wires were common. Programs were stored on cassette tape cartridges. Facilities for printing and documentation were minimal due to lack of memory capacity. The oldest PLCs used non-volatile magnetic core memory.

More recently, PLCs are programmed using application software on personal computers, which now represent the logic in graphic form instead of character symbols. The computer is connected to the PLC through USB, Ethernet, RS-232, RS-485, or RS-422 cabling. The programming software allows entry and editing of the ladder-style logic. In some software packages, it is also possible to view and edit the program in function block diagrams, sequence flow charts and structured text. Generally the software provides functions for debugging and troubleshooting the PLC software, for example, by highlighting portions of the logic to show current status during operation or via simulation. The software will upload and download the PLC program, for backup and restoration purposes. In some models of programmable controller, the program is transferred from a personal computer to the PLC through a programming board which writes the program into a removable chip such as an EPROM.

Under the IEC 61131-3 standard, PLCs can be programmed using standards-based programming languages. The most commonly used programming language is Ladder diagram (LD) also known as Ladder logic. It uses Contact-Coil logic to make programs like an electrical control diagram. A graphical programming notation called Sequential

Function Charts is available on certain programmable controllers. A model which emulated electromechanical control panel devices (such as the contact and coils of relays) which PLCs replaced. This model remains common today.

IEC 61131-3 currently defines four programming languages for programmable control systems: function block diagram (FBD), ladder diagram (LD), structured text (ST; similar to the Pascal programming language), and sequential function chart (SFC). Instruction list (IL; similar to assembly language) was deprecated in the third edition of the standard. These techniques emphasize logical organization of operations.

While the fundamental concepts of PLC programming are common to all manufacturers, differences in I/O addressing, memory organization, and instruction sets mean that PLC programs are never perfectly interchangeable between different makers. Even within the same product line of a single manufacturer, different models may not be directly compatible.

Control Example Shown in Ladder Diagram

This is a programming example in ladder diagram which shows the control system. A ladder diagram is a method of drawing control circuits which pre-dates PLCs. The ladder diagram resembles the schematic diagram of a system built with electromechanical relays.

As an example, say a facility needs to store water in a tank. The water is drawn from the tank by another system, as needed, and our example system must manage the water level in the tank by controlling the valve that refills the tank. Shown are:

- Two inputs (from the low and high level switches) represented by contacts of the float switches.
- An output to the fill valve, labelled as the fill valve which it controls.
- An "internal" contact, representing the output signal to the fill valve which is created in the program.
- A logical control scheme created by the interconnection of these items in software.

In ladder diagram, the contact symbols represent the state of bits in processor memory, which corresponds to the state of physical inputs to the system. If a discrete input is energized, the memory bit is a 1, and a "normally open" contact controlled by that bit will pass a logic "true" signal on to the next element of the ladder. Therefore, the contacts in the PLC program that "read" or look at the physical switch contacts in this case must be "opposite" or open in order to return a TRUE for the closed physical switches. Internal status bits, corresponding to the state of discrete outputs, are also available to the program.

In the example, the physical state of the float switch contacts must be considered when choosing "normally open" or "normally closed" symbols in the ladder diagram. The PLC has two discrete inputs from float switches (Low Level and High Level). Both float switches (normally closed) open their contacts when the water level in the tank is above the physical location of the switch.

When the water level is below both switches, the float switch physical contacts are both closed, and a true (logic 1) value is passed to the Fill Valve output. Water begins to fill the tank. The internal "Fill Valve" contact latches the circuit so that even when the "Low Level" contact opens (as the water passes the lower switch), the fill valve remains on. Since the High Level is also normally closed, water continues to flow as the water level remains between the two switch levels. Once the water level rises enough so that the "High Level" switch is off (opened), the PLC will shut the inlet to stop the water from overflowing; this is an example of seal-in (latching) logic. The output is sealed in until a high level condition breaks the circuit. After that the fill valve remains off until the level drops so low that the Low Level switch is activated, and the process repeats again.

```
| (N.C. physical  (N.C. physical                               |
|    Switch)         Switch)                                   |
|   Low Level       High Level               Fill Valve        |
|---[ ]---|---[ ]-----------(OUT)-----|
|             |                                                |
|             |                                                |
|             |                                                |
|  Fill Valve |                                                |
|---[ ]---|                                               |
|                                                              |
|                                                              |
```

A complete program may contain thousands of rungs, evaluated in sequence. Typically the PLC processor will alternately scan all its inputs and update outputs, then evaluate the ladder logic; input changes during a program scan will not be effective until the next I/O update. A complete program scan may take only a few milliseconds, much faster than changes in the controlled process.

Programmable controllers vary in their capabilities for a "rung" of a ladder diagram. Some only allow a single output bit. There are typically limits to the number of series contacts in line, and the number of branches that can be used. Each element of the rung is evaluated sequentially. If elements change their state during evaluation of a rung, hard-to-diagnose faults can be generated, although sometimes (as above) the technique is useful. Some implementations forced evaluation from left-to-right as displayed and did not allow reverse flow of a logic signal (in multi-branched rungs) to affect the output.

Functionality

The functionality of the PLC has evolved over the years to include sequential relay control, motion control, process control, distributed control systems, and networking. The data handling, storage, processing power, and communication capabilities of some modern PLCs are approximately equivalent to desktop computers. PLC-like programming combined with remote I/O hardware, allow a general-purpose desktop computer to overlap some PLCs in certain applications. Desktop computer controllers have not been generally accepted in heavy industry because the desktop computers run on less stable operating systems than PLCs, and because the desktop computer hardware is typically not designed to the same levels of tolerance to temperature, humidity, vibration, and longevity as the processors used in PLCs. Operating systems such as Windows do not lend themselves to deterministic logic execution, with the result that the controller may not always respond to changes of input status with the consistency in timing expected from PLCs. Desktop logic applications find use in less critical situations, such as laboratory automation and use in small facilities where the application is less demanding and critical.

Basic Functions

The most basic function of a programmable controller is to emulate the functions of electro-mechanical relays. Discrete inputs are given a unique address, and a PLC instruction can test if the input state is on or off. Just as a series of relay contacts perform a logical AND function, not allowing current to pass unless all the contacts are closed, so a series of "examine if on" instructions will energize its output storage bit if all the input bits are on. Similarly, a parallel set of instructions will perform a logical OR. In an electro-mechanical relay wiring diagram, a group of contacts controlling one coil is called a "rung" of a "ladder diagram ", and this concept is also used to describe PLC logic. Some models of PLC limit the number of series and parallel instructions in one "rung" of logic. The output of each rung sets or clears a storage bit, which may be associated with a physical output address or which may be an "internal coil" with no physical connection. Such internal coils can be used, for example, as a common element in multiple separate rungs. Unlike physical relays, there is usually no limit to the number of times an input, output or internal coil can be referenced in a PLC program.

Some PLCs enforce a strict left-to-right, top-to-bottom execution order for evaluating the rung logic. This is different from electro-mechanical relay contacts, which in a sufficiently complex circuit may either pass current left-to-right or right-to-left, depending on the configuration of surrounding contacts. The elimination of these "sneak paths" is either a bug or a feature, depending on programming style.

More advanced instructions of the PLC may be implemented as functional blocks, which carry out some operation when enabled by a logical input and which produce outputs to signal, for example, completion or errors, while manipulating variable internally that may not correspond to discrete logic.

Timers and Counters

The main function of a timer is to keep an output on for a specific length of time. A good example of this is a garage light, where you want power to be cut off after 2 minutes so as to give someone time to go into the house. The three different types of timers that are commonly used are a Delay-OFF, a Delay-ON, and a Delay-ON-Retentive. A Delay-OFF timer activates immediately when turned on, counts down from a programmed time before cutting off, and is cleared when the enabling input is off. A Delay-ON timer is activated by input and starts accumulating time, counts up to a programmed time before cutting off, and is cleared when the enabling input is turned off. A Delay-ON-Retentive timer is activated by input and starts accumulating time, retains the accumulated value even if the (ladder-logic) rung goes false, and can be reset only by a RESET contact.

Counters are primarily used for counting items such as cans going into a box on an assembly line. This is important because once something is filled to its max the item needs to be moved on so something else can be filled. Many companies use counters in PLC's to count boxes, count how many feet of something is covered, or to count how many pallets are on a truck. There are three types of counters, Up counters, Down counters, and Up/Down counters. Up counters count up to the preset value, turn on the CTU (CounT Up output) when the preset value is reached, and are cleared upon receiving a reset. Down counters count down from a preset value, turns on the CTD (CounT Down output) when 0 is reached, and are cleared upon reset. Up/Down counters count up on CU, count down on CD, turn on CTUD (CounT Up/Down output) when the preset value is reached, and cleared on reset.

Programmable Logic Relay

In more recent years, small products called PLRs (programmable logic relays), and also by similar names, have become more common and accepted. These are much like PLCs, and are used in light industry where only a few points of I/O (i.e. a few signals coming in from the real world and a few going out) are needed, and low cost is desired. These small devices are typically made in a common physical size and shape by several manufacturers, and branded by the makers of larger PLCs to fill out their low end product range. Popular names include PICO Controller, NANO PLC, and other names implying very small controllers. Most of these have 8 to 12 discrete inputs, 4 to 8 discrete outputs, and up to 2 analog inputs. Size is usually about 4" wide, 3" high, and 3" deep. Most such devices include a tiny postage-stamp-sized LCD screen for viewing simplified ladder logic (only a very small portion of the program being visible at a given time) and status of I/O points, and typically these screens are accompanied by a 4-way rocker push-button plus four more separate push-buttons, similar to the key buttons on a VCR remote control, and used to navigate and edit the logic. Most have a small plug for connecting via RS-232 or RS-485 to a personal computer so that programmers can use simple Windows applications for programming instead of being forced to use the tiny LCD and push-button set for this purpose. Unlike regular PLCs that are usually modular and greatly expandable, the PLRs are usually not modular or expandable, but their price can be two orders of magnitude less than a PLC, and they still offer robust design and deterministic execution of the logics.

PLC Topics

Features

The main difference from most other computing devices is that PLCs are intended-for and therefore tolerant-of more severe conditions (such as dust, moisture, heat, cold), while offering extensive input/output (I/O) to connect the PLC to sensors and actuators. PLC input can include simple digital elements such as limit switches, analog variables from process sensors (such as temperature and pressure), and more complex data such as that from positioning or machine vision systems. PLC output can include elements such as indicator lamps, sirens, electric motors, pneumatic or hydraulic cylinders, magnetic relays, solenoids, or analog outputs. The input/output arrangements may be built into a simple PLC, or the PLC may have external I/O modules attached to a fieldbus or computer network that plugs into the PLC.

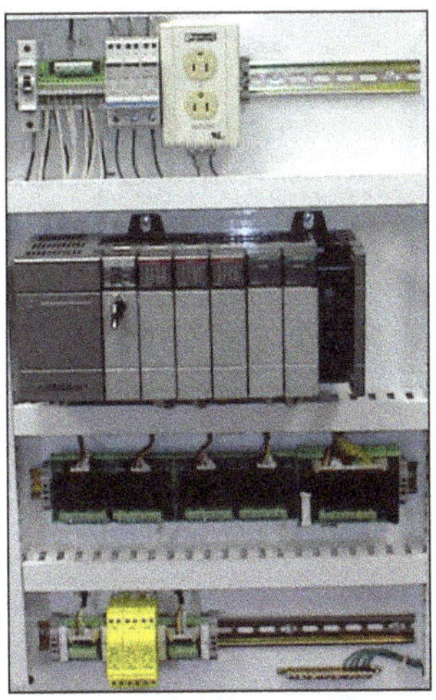

Control panel with PLC (grey elements in the center).
The unit consists of separate elements, from left to right;
power supply, controller, relay units for in- and output.

Scan Time

A PLC program generally loops i.e. executes repeatedly, as long as the controlled system is running. At the start of each execution loop, the status of all physical inputs are copied to an area of memory, sometimes called the "I/O Image Table", which is accessible to the processor. The program then runs from its first instruction rung down to the last rung. It takes some time for the processor of the PLC to evaluate all the rungs and update the I/O image table with the status of outputs. Scan times of a few milliseconds may be encountered for small programs and fast processors, but for older processors and very large programs much longer scan times (on the order of 100 ms) may be encountered. Excessively long scan times may mean the response of the PLC to changing inputs or process conditions is too slow to be useful.

Control panel with a PLC user interface for thermal oxidizer regulation.

As PLCs became more advanced, methods were developed to change the sequence of ladder execution, and subroutines were implemented. This enhanced programming could be used to save scan time for high-speed processes; for example, parts of the program used only for setting up the machine could be segregated from those parts required to operate at higher speed. Newer PLCs now have the option to run the logic program synchronously with the IO scanning. This means that IO is updated in the background and the logic reads and writes values as required during the logic scanning.

Special-purpose I/O modules may be used where the scan time of the PLC is too long to allow predictable performance. Precision timing modules, or counter modules for use with shaft encoders, are used where the scan time would be too long to reliably count pulses or detect the sense of rotation of an encoder. This allows even a relatively slow PLC to still interpret the counted values to control a machine, as the accumulation of pulses is done by a dedicated module that is unaffected by the speed of program execution on the PLC.

Process of a Scan Cycle

There are five main steps in a scan cycle:

- Reading inputs,
- Executing the program,

- Processing communication requests,
- Executing CPU diagnostics,
- Writing outputs.

System Scale

A small PLC will have a fixed number of connections built in for inputs and outputs. Typically, expansions are available if the base model has insufficient I/O.

Modular PLCs have a chassis (also called a rack) into which are placed modules with different functions. The processor and selection of I/O modules are customized for the particular application. Several racks can be administered by a single processor, and may have thousands of inputs and outputs. Either a special high speed serial I/O link or comparable communication method is used so that racks can be distributed away from the processor, reducing the wiring costs for large plants. Options are also available to mount I/O points directly to the machine and utilize quick disconnecting cables to sensors and valves, saving time for wiring and replacing components.

User Interface

PLCs may need to interact with people for the purpose of configuration, alarm reporting, or everyday control. A human-machine interface (HMI) is employed for this purpose. HMIs are also referred to as man-machine interfaces (MMIs) and graphical user interfaces (GUIs). A simple system may use buttons and lights to interact with the user. Text displays are available as well as graphical touch screens. More complex systems use programming and monitoring software installed on a computer, with the PLC connected via a communication interface.

Communications

Many models of PLCs have built-in communications ports, using RS-232, RS-422, RS-485, or Ethernet. Various protocols are usually included. Many of these protocols are vendor specific. Most modern PLCs can communicate over a network to some other system, such as a computer running a SCADA (Supervisory Control And Data Acquisition) system or web browser.

PLCs used in larger I/O systems may have peer-to-peer (P2P) communication between processors. This allows separate parts of a complex process to have individual control while allowing the subsystems to co-ordinate over the communication link. These communication links are also often used for HMI devices such as keypads or PC-type workstations.

Formerly, some manufacturers offered dedicated communication modules as an add-on function where the processor had no network connection built-in.

Security

Prior to the discovery of the Stuxnet computer worm in June 2010, security of PLCs received little attention. Modern PLCs generally contain a real-time operating system such as OS-9 or VxWorks, and exploits for these systems exist much as they do for desktop computer operating systems such as Microsoft Windows. PLCs can also be attacked by gaining control of a computer they communicate with.

Simulation

In order to properly understand the operation of a PLC, it is necessary to spend considerable time programming, testing, and debugging PLC programs. PLC systems are inherently expensive, and down-time is often very costly. In addition, if a PLC is programmed incorrectly it can result in lost productivity and dangerous conditions. PLC simulation software such as PLCLogix can save time in the design of automated control applications and can also increase the level of safety associated with equipment since many "what if" scenarios can be tried and tested before the system is activated.

Redundancy

Some special processes need to work permanently with minimum unwanted down time. Therefore, it is necessary to design a system which is fault-tolerant and capable of handling the process with faulty modules. In such cases to increase the system availability in the event of hardware component failure, redundant CPU or I/O modules with the same functionality can be added to hardware configuration for preventing total or partial process shutdown due to hardware failure.

PLC Compared with other Control Systems

PLC installed in a control panel.

PLCs are well adapted to a range of automation tasks. These are typically industrial processes in manufacturing where the cost of developing and maintaining the automation system is high relative to the total cost of the automation, and where changes to the system would be expected during its operational life. PLCs contain input and output devices compatible with industrial pilot devices and controls; little electrical design is required, and the design problem centers on expressing the desired sequence of operations. PLC applications are typically highly customized systems, so the cost of a packaged PLC is low compared to the cost of a specific custom-built controller design. On the other hand, in the case of mass-produced goods, customized control systems are economical. This is due to the lower cost of the components, which can be optimally chosen instead of a "generic" solution, and where the non-recurring engineering charges are spread over thousands or millions of units.

Control center with a PLC for a RTO.

For high volume or very simple fixed automation tasks, different techniques are used. For example, a cheap consumer dishwasher would be controlled by an electromechanical cam timer costing only a few dollars in production quantities.

A microcontroller-based design would be appropriate where hundreds or thousands of units will be produced and so the development cost (design of power supplies, input/output hardware, and necessary testing and certification) can be spread over many sales, and where the end-user would not need to alter the control. Automotive applications are an example; millions of units are built each year, and very few end-users alter the programming of these controllers. However, some specialty vehicles such as transit buses economically use PLCs instead of custom-designed controls, because the volumes are low and the development cost would be uneconomical.

Very complex process control, such as used in the chemical industry, may require algorithms and performance beyond the capability of even high-performance PLCs. Very high-speed or precision controls may also require customized solutions; for example, aircraft flight controls. Single-board computers using semi-customized or fully proprietary hardware may be chosen for very demanding control applications where the high development and maintenance cost can be supported. "Soft PLCs" running on desktop-type computers can interface with industrial I/O hardware while executing programs within a version of commercial operating systems adapted for process control needs.

Programmable controllers are widely used in motion, positioning, or torque control. Some manufacturers produce motion control units to be integrated with PLC so that G-code (involving a CNC machine) can be used to instruct machine movements.

PLCs may include logic for single-variable feedback analog control loop, a proportional, integral, derivative (PID) controller. A PID loop could be used to control the temperature of a manufacturing process, for example. Historically PLCs were usually configured with only a few analog control loops; where processes required hundreds or thousands of loops, a distributed control system (DCS) would instead be used. As PLCs have become more powerful, the boundary between DCS and PLC applications has been blurred.

PLCs have similar functionality as remote terminal units. An RTU, however, usually does not support control algorithms or control loops. As hardware rapidly becomes more powerful and cheaper, RTUs, PLCs, and DCSs are increasingly beginning to overlap in responsibilities, and many vendors sell RTUs with PLC-like features, and vice versa. The industry has standardized on the IEC 61131-3 functional block language for creating programs to run on RTUs and PLCs, although nearly all vendors also offer proprietary alternatives and associated development environments.

In recent years "safety" PLCs have started to become popular, either as standalone models or as functionality and safety-rated hardware added to existing controller architectures (Allen-Bradley Guardlogix, Siemens F-series etc.). These differ from conventional PLC types as being suitable for use in safety-critical applications for which PLCs have traditionally been supplemented with hard-wired safety relays. For example, a safety PLC might be used to control access to a robot cell with trapped-key access, or perhaps to manage the shutdown response to an emergency stop on a conveyor production line. Such PLCs typically have a restricted regular instruction set augmented with safety-specific instructions designed to interface with emergency stops, light screens, and so forth. The flexibility that such systems offer has resulted in rapid growth of demand for these controllers.

The rising popularity of single board computers has also had an influence on the development of PLCs. Traditional PLCs are generally closed platforms, but some newer PLCs (e.g. ctrlX from Bosch Rexroth, PFC200 from Wago, PLCnext from Phoenix Contact, and Revolution Pi from Kunbus) provide the features of traditional PLCs on an open platform.

Discrete (Digital) and Analog Signals

Discrete (digital) signals behave as binary switches, yielding simply an On or Off signal (1 or 0, True or False, respectively). Push buttons, limit switches, and photoelectric sensors are examples of devices providing a discrete signal. Discrete signals are sent using either voltage or current, where a specific range is designated as On and another as Off. For example, a PLC might use 24 V DC I/O, with values above 22 V DC representing On, values below 2VDC representing *Off*, and intermediate values undefined. Initially, PLCs had only digital I/O.

Analog signals are like volume controls, with a range of values between zero and full-scale. These are typically interpreted as integer values (counts) by the PLC, with various ranges of accuracy depending on the device and the number of bits available to store the data. As PLCs typically use 16-bit signed binary processors, the integer values are limited between -32,768 and +32,767. Pressure, temperature, flow, and weight are often represented by analog signals. Analog signals can use voltage or current with a magnitude proportional to the value of the process signal. For example, an analog 0 to 10 V or 4-20 mA input would be converted into an integer value of 0 to 32767.

Current inputs are less sensitive to electrical noise (e.g. from welders or electric motor starts) than voltage inputs.

PLCs are at the forefront of manufacturing automation. An engineer working in a manufacturing environment will at least encounter some PLCs, if not use them on a regular basis. Electrical engineering students should have basic knowledge of PLCs because of their widespread use in industrial applications.

Ladder Logic

Ladder logic was originally a written method to document the design and construction of relay racks as used in manufacturing and process control. Each device in the relay rack would be represented by a symbol on the ladder diagram with connections between those devices shown. In addition, other items external to the relay rack such as pumps, heaters, and so forth would also be shown on the ladder diagram.

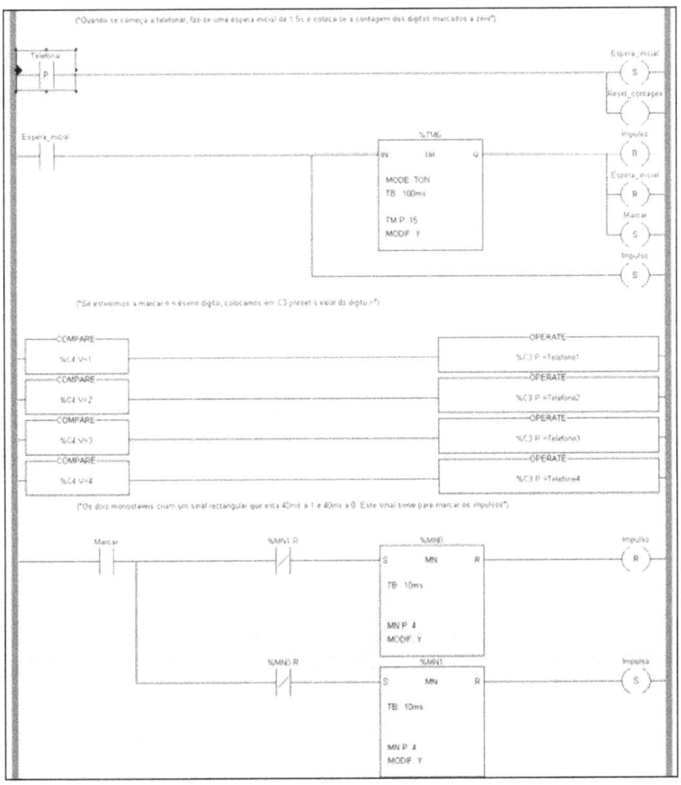

Part of a ladder diagram, including contacts and coils, compares, timers and monostable multivibrators.

Ladder logic has evolved into a programming language that represents a program by a graphical diagram based on the circuit diagrams of relay logic hardware. Ladder logic is used to develop software for programmable logic controllers (PLCs) used in industrial control applications. The name is based on the observation that programs in this language resemble ladders, with two vertical rails and a series of horizontal rungs between them. While ladder diagrams were once the only available notation for recording programmable controller programs, today other forms are standardized in IEC 61131-3 (For example, as an alternative to the graphical ladder logic form, there is also a more assembly language like format called Instruction list within the IEC 61131-3 standard.).

Ladder logic is widely used to program PLCs, where sequential control of a process or manufacturing operation is required. Ladder logic is useful for simple but critical control systems or for reworking old hardwired relay circuits. As programmable logic controllers became more sophisticated it has also been used in very complex automation systems. Often the ladder logic program is used in conjunction with an HMI program operating on a computer workstation.

The motivation for representing sequential control logic in a ladder diagram was to allow factory engineers and technicians to develop software without additional training to learn a language such as FORTRAN or other general purpose computer language. Development and maintenance were simplified because of the resemblance to familiar relay hardware systems. Implementations of ladder logic may have characteristics, such as sequential execution and support for control flow features, that make the analogy to hardware somewhat inaccurate.

Ladder logic can be thought of as a rule-based language rather than a procedural language. A "rung" in the ladder represents a rule. When implemented with relays and other electromechanical devices, the various rules execute simultaneously and immediately. When implemented in a programmable logic controller, the rules are typically executed sequentially by software in a continuous loop, or "scan". By executing the loop fast enough, typically many times per second, the effect of simultaneous and immediate execution is achieved. Proper use of programmable controllers requires an understanding of the limitations of the execution order of rungs.

Syntax and Examples

The language itself can be seen as a set of connections between logical checkers (contacts) and actuators (coils). If a path can be traced between the left side of the rung and the output, through asserted (true or "closed") contacts, the rung is true and the output coil storage bit is asserted (1) or true. If no path can be traced, then the output is false (0) and the "coil" by analogy to electromechanical relays is considered "de-energized". The analogy between logical propositions and relay contact status is due to Claude Shannon.

Ladder logic has contacts that make or break circuits to control coils. Each coil or contact corresponds to the status of a single bit in the programmable controller's memory. Unlike electromechanical relays, a ladder program can refer any number of times to the status of a single bit, equivalent to a relay with an indefinitely large number of contacts.

So-called "contacts" may refer to physical ("hard") inputs to the programmable controller from physical devices such as pushbuttons and limit switches via an integrated or external input module, or may represent the status of internal storage bits which may be generated elsewhere in the program.

Each rung of ladder language typically has one coil at the far right. Some manufacturers may allow more than one output coil on a rung.

- Rung input : checkers (contacts):
    - —[ ]— Normally open contact, closed whenever its corresponding coil or an input which controls it is energized. (Open contact at rest).
    - —[\]— Normally closed ("not") contact, closed whenever its corresponding coil or an input which controls it is not energized. (Closed contact at rest).
- Rung output: actuators (coils):
    - —( )— Normally inactive coil, energized whenever its rung is closed. (Inactive at rest).
    - —(\)— Normally active ("not") coil, energized whenever its rung is open. (Active at rest).

The "coil" (output of a rung) may represent a physical output which operates some device connected to the programmable controller, or may represent an internal storage bit for use elsewhere in the program.

A way to recall these is to imagine the checkers (contacts) as a push button input, and the actuators (coils) as a light bulb output. The presence of a slash within the checkers or actuators would indicate the default state of the device at rest.

Logical AND

```
---[ ]-------[ ]--------( )

  Key switch 1     Key switch 2        Door motor
```

The above realizes the function: Door motor = Key switch 1 AND Key switch 2.

This circuit shows two key switches that security guards might use to activate an electric motor on a bank vault door. When the normally open contacts of both switches close, electricity is able to flow to the motor which opens the door.

Logical AND with NOT

```
---[ ]-------[\]--------( )
   Close door      Obstruction      Door motor
```

The above realizes the function: Door motor = Close door AND NOT(Obstruction).

This circuit shows a push button that closes a door, and an obstruction detector that senses if something is in the way of the closing door. When the normally open push button contact closes and the normally closed obstruction detector is closed (no obstruction detected), electricity is able to flow to the motor which closes the door.

Logical OR

```
-+----[ ]----+---------( )
 | Exterior unlock |            Unlock
 |                 |
 +----[ ]----+
   Interior unlock
```

The above realizes the function: Unlock = Interior unlock OR Exterior unlock.

This circuit shows the two things that can trigger a car's power door locks. The remote receiver is always powered. The unlock solenoid gets power when either set of contacts is closed.

Industrial Stop/Start

In common industrial latching start/stop logic we have a "Start" button to turn on a motor contactor, and a "Stop" button to turn off the contactor.

When the "Start" button is pushed the input goes true, via the "Stop" button NC contact. When the "Run" input becomes true the seal-in "Run" NO contact in parallel with the "Start" NO contact will close maintaining the input logic true (latched or sealed-in). After the circuit is latched the "Stop" button may be pushed causing its NC contact to open and consequently the input to go false. The "Run" NO contact then opens and the circuit logic returns to its inactive state.

```
-+--[ ]-+--[\]--( )
 |  Start |   Stop  Run
 |        |
 +--[ ]-+
```

```
        Run
----[ ]-------( )
        Run                Motor
```

The above realizes the function: Run = (Start OR Run) AND (NOT Stop).

This latch configuration is a common idiom in ladder logic. It may also be referred to as "seal-in logic". The key to understanding the latch is in recognizing that the "Start" switch is a momentary switch (once the user releases the button, the switch is open again). As soon as the "Run" solenoid engages, it closes the "Run" NO contact, which latches the solenoid on. The "Start" switch opening up then has no effect.

In this example, "Run" represents the status of a bit in the PLC, while "Motor" represents the actual output to the real-world relay that closes the motor's real-world circuit.

For safety reasons, an Emergency-Stop may be hardwired in series with the Start switch, and the relay logic should reflect this.

```
-[\]--[\]--+-[ ]-+-----( )
  ES    Stop   | Start |        Run
               |       |
               +-[ ]-+
                  Run
----[ ]-------( )
        Run                Motor
```

The above realizes the function: Run = (NOT Emergency Stop) AND (NOT Stop) AND (Start OR Run).

Complex Logic

Here is an example of what two rungs in a ladder logic program might look like. In real world applications, there may be hundreds or thousands of rungs.

Typically, complex ladder logic is 'read' left to right and top to bottom. As each of the lines (or rungs) are evaluated the output coil of a rung may feed into the next stage of the ladder as an input. In a complex system there will be many "rungs" on a ladder, which are numbered in order of evaluation.

```
1. --[ ]-----+--[ ]---+--( )
     Switch      |   HiTemp   |    A/C
                 |            |
```

```
                        +--[ ]---+

                            Humid

2. --[ ]--[\]----------( )

       A/C    Heat                  Cooling
```

Line 1 realizes the function: A/C = Switch AND (HiTemp OR Humid).

Line 2 realizes the function: Cooling = A/C AND (NOT Heat).

This represents a slightly more complex system for rung 2. After the first line has been evaluated, the output coil "A/C" is fed into rung 2, which is then evaluated and the output coil "Cooling" could be fed into an output device "Compressor" or into rung 3 on the ladder. This system allows very complex logic designs to be broken down and evaluated.

Additional Functionality

Additional functionality can be added to a ladder logic implementation by the PLC manufacturer as a special block. When the special block is powered, it executes code on predetermined arguments. These arguments may be displayed within the special block.

```
                                  +-----+

---[ ]----------+   A    +--

Remote unlock                     +-----+

                                Remote counter

                                  +-----+

---[ ]----------+   B    +--

Interior unlock                   +-----+

                               Interior counter

                          +-----+

----------+  A + B +------

                          | into C |

                          +-----+

                            Adder
```

In this example, the system will count the number of times that the interior and remote unlock buttons are pressed. This information will be stored in memory locations A and B. Memory location C will hold the total number of times that the door has been unlocked electronically.

PLCs have many types of special blocks. They include timers, arithmetic operators and comparisons, table lookups, text processing, PID control, and filtering functions. More powerful PLCs can operate on a group of internal memory locations and execute an operation on a range of addresses, for example, to simulate a physical sequential drum controller or a finite state machine. In some cases, users can define their own special blocks, which effectively are subroutines or macros. The large library of special blocks along with high speed execution has allowed use of PLCs to implement very complex automation systems.

Limitations and Successor Languages

Ladder notation is best suited to control problems where only binary variables are required and where interlocking and sequencing of binary is the primary control problem. Like all parallel programming languages, the sequential order of operations may be undefined or obscure; logic race conditions are possible which may produce unexpected results. Complex rungs are best broken into several simpler steps to avoid this problem. Some manufacturers avoid this problem by explicitly and completely defining the execution order of a rung, however programmers may still have problems fully grasping the resulting complex semantics.

Analog quantities and arithmetical operations are clumsy to express in ladder logic and each manufacturer has different ways of extending the notation for these problems. There is usually limited support for arrays and loops, often resulting in duplication of code to express cases which in other languages would call for use of indexed variables.

As microprocessors have become more powerful, notations such as sequential function charts and function block diagrams can replace ladder logic for some limited applications. Some newer PLCs may have all or part of the programming carried out in a dialect that resembles BASIC, C, or other programming language with bindings appropriate for a real-time application environment.

Structured Text

Structured text, abbreviated as ST or STX, is one of the five languages supported by the IEC 61131-3 standard, designed for programmable logic controllers (PLCs). It is a high level language that is block structured and syntactically resembles Pascal, on which it is based. All of the languages share IEC61131 Common Elements. The variables and function calls are defined by the common elements so different languages within the IEC 61131-3 standard can be used in the same program.

Complex statements and nested instructions are supported:

- Iteration loops (REPEAT-UNTIL; WHILE-DO),
- Conditional execution (IF-THEN-ELSE; CASE),
- Functions (SQRT(), SIN()).

Sample Program

```
(* simple state machine *)

TxtState := STATES[StateMachine];

CASE StateMachine OF

   1: ClosingValve();

      StateMachine := 2;

   2: OpeningValve();

ELSE

    BadCase();

END_CASE;
```

Unlike in some other programming languages, there is no fallthrough for the CASE statement: the first matching condition is entered, and after running its statements, the CASE block is left without checking other conditions.

Additional ST Programming Examples

```
//PLC configuration

CONFIGURATION DefaultCfg

    VAR_GLOBAL

        b_Start_Stop : BOOL;    //Global variable to represent a bool-
ean.

         b_ON_OFF : BOOL;      //Global variable to represent a boolean.

        Start_Stop AT %IX0.0:BOOL; //Digital input of the PLC (Address
0.0)

     ON_OFF AT %QX0.0:BOOL; //Digital output of the PLC (Address 0.0).
(Coil)

   END_VAR

   //Schedule the main program to be executed every 20 ms

   TASK Tick(INTERVAL := t20ms);

   PROGRAM Main WITH Tick : Monitor_Start_Stop;

END_CONFIGURATION

PROGRAM Monitor_Start_Stop            //Actual Program

   VAR_EXTERNAL

       Start_Stop : BOOL;
```

```
        ON_OFF : BOOL;
    END_VAR
    VAR                             //Temporary variables for logic han-
dling
        ONS_Trig : BOOL;
        Rising_ONS : BOOL;
    END_VAR
    //Start of Logic
    //Catch the Rising Edge One Shot of the Start_Stop input
    ONS_Trig := Start_Stop AND NOT Rising_ONS;

    //Main Logic for Run_Contact - Toggle ON /Toggle OFF --
    ON_OFF := (ONS_Trig AND NOT ON_OFF) OR (ON_OFF AND NOT ONS_Trig);
    //Rising One Shot logic
    Rising_ONS := Start_Stop;
END_PROGRAM
```

Function Block Example

```
//=================================================================
//Function Block Timed Counter : Incremental count of the timed interval
//=================================================================
FUNCTION_BLOCK FB_Timed_Counter
    VAR_INPUT
        Execute : BOOL := FALSE;        //Trigger signal to begin Timed
Counting
          Time_Increment : REAL := 1.25; //Enter Cycle Time (Seconds)
between counts
          Count_Cycles : INT := 20; //Number of Desired Count Cycles
    END_VAR

    VAR_OUTPUT
        Timer_Done_Bit : BOOL := FALSE; //One Shot Bit indicating Timer
Cycle Done
```

```
    Count_Complete : BOOL := FALSE; //Output Bit indicating the Count
is complete
        Current_Count : INT := 0; //Accumulating Value of Counter
    END_VAR

    VAR
        CycleTimer : TON; //Timer FB from Command Library
        CycleCounter : CTU; //Counter FB from Command Library
        TimerPreset : TIME; //Converted Time_Increment in Seconds to MS
    END_VAR

    //Start of Function Block programming
    TimerPreset := REAL_TO_TIME(in := Time_Increment) * 1000;
    CycleTimer(
         in := Execute AND NOT CycleTimer.Q,
         pt := TimerPreset);
    Timer_Done_Bit := CycleTimer.Q;

    CycleCounter(
        cu := CycleTimer.Q,
        r := NOT Execute,
      pv := Count_Cycles);
    Current_Count := CycleCounter.cv;
    Count_Complete := CycleCounter.q;

END_FUNCTION_BLOCK
```

Instruction List

Instruction List (IL) is one of the 5 languages supported by the initial versions of IEC 61131-3 standard. It is designed for programmable logic controllers (PLCs). It is a low level language and resembles assembly. All of the languages share IEC61131 Common Elements. The variables and function call are defined by the common elements so different languages can be used in the same program.

Program control (control flow) is achieved by *jump* instructions and function calls (subroutines with optional parameters).

The file format has now been standardized to XML by PLCopen.

Example:

```
          LD     Speed
          GT     2000
          JMPCN  VOLTS_OK
          LD     Volts
VOLTS_OK  LD     1
          ST     %Q75
```

Variations from IEC 61131

Many vendors whilst incorporating the full IEC 61131-3 requirements have additional vendor specific calls/function blocks to suit their hardware such as reading or writing to I/O. Siemens PLC instruction list language is known as "Statement List" or "STL" in English, and "Anweisungs-Liste" or "AWL" in German, Italian and Spanish. The user of a Simatic development package may choose between German and International mnemonics to represent instructions. For example, "A" for "AND" or "U" for "UND", "I" for "Input" or "E" for "Eingang".

Supervisory Control and Data Acquisition

Supervisory control and data acquisition (SCADA) is a control system architecture comprising computers, networked data communications and graphical user interfaces(GUI) for high-level process supervisory management, while also comprising other peripheral devices like programmable logic controllers (PLC) and discrete proportional-integral-derivative (PID) controllers to interface with process plant or machinery. The use of SCADA has been considered also for management and operations of project-driven-process in construction.

The operator interfaces which enable monitoring and the issuing of process commands, like controller set point changes, are handled through the SCADA computer system. The subordinated operations, e.g. the real-time control logic or controller calculations, are performed by networked modules connected to the field sensors and actuators.

The SCADA concept was developed to be a universal means of remote-access to a variety of local control modules, which could be from different manufacturers and allowing access through standard automation protocols. In practice, large SCADA systems have grown to become very similar to distributed control systems in function, while using multiple means of interfacing with the plant. They can control large-scale processes that can include multiple sites, and work over large distances as well as small distance. It is one of the most commonly-used types of industrial control systems, in spite of concerns about SCADA systems being vulnerable to cyberwarfare/cyberterrorism attacks.

The SCADA Concept in Control Operations

The key attribute of a SCADA system is its ability to perform a supervisory operation over a variety of other proprietary devices. The accompanying diagram is a general model which shows functional manufacturing levels using computerised control.

Referring to the diagram:

- Level 0 contains the field devices such as flow and temperature sensors, and final control elements, such as control valves.
- Level 1 contains the industrialised input/output (I/O) modules, and their associated distributed electronic processors.
- Level 2 contains the supervisory computers, which collate information from processor nodes on the system, and provide the operator control screens.
- Level 3 is the production control level, which does not directly control the process, but is concerned with monitoring production and targets.
- Level 4 is the production scheduling level.

Level 1 contains the programmable logic controllers (PLCs) or remote terminal units (RTUs).

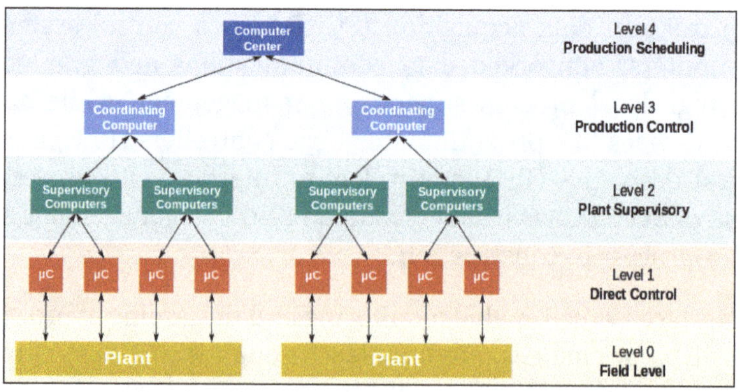

Functional levels of a manufacturing control operation.

Level 2 contains the SCADA to readings and equipment status reports that are communicated to level 2 SCADA as required. Data is then compiled and formatted in such a way that a control room operator using the HMI (Human Machine Interface) can make supervisory decisions to adjust or override normal RTU (PLC) controls. Data may also be fed to a historian, often built on a commodity database management system, to allow trending and other analytical auditing.

SCADA systems typically use a tag database, which contains data elements called tags or points, which relate to specific instrumentation or actuators within the process system. Data is accumulated against these unique process control equipment tag references.

Examples of Use

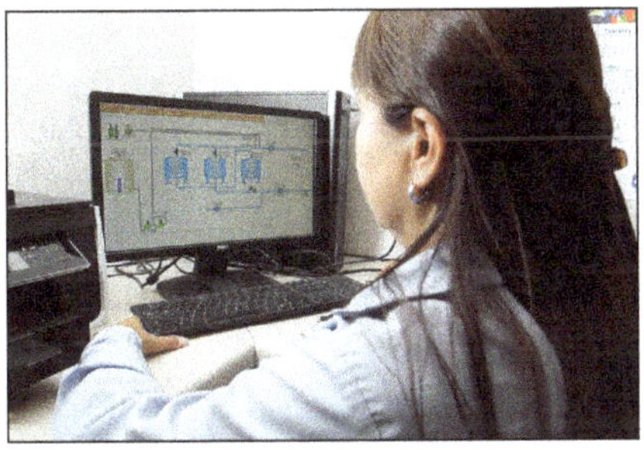

Example of SCADA used in office environment to remotely monitor a process.

Both large and small systems can be built using the SCADA concept. These systems can range from just tens to thousands of control loops, depending on the application. Example processes include industrial, infrastructure, and facility-based processes:

- Industrial processes include manufacturing, Process control, power generation, fabrication, and refining, and may run in continuous, batch, repetitive, or discrete modes.
- Infrastructure processes may be public or private, and include water treatment and distribution, wastewater collection and treatment, oil and gas pipelines, electric power transmission and distribution, and wind farms.
- Facility processes, including buildings, airports, ships, and space stations. They monitor and control heating, ventilation, and air conditioning systems (HVAC), access, and energy consumption.

However, SCADA systems may have security vulnerabilities, so the systems should be evaluated to identify risks and solutions implemented to mitigate those risks.

SCADA System Components

A SCADA system usually consists of the following main elements:

Supervisory Computers

This is the core of the SCADA system, gathering data on the process and sending control commands to the field connected devices. It refers to the computer and software responsible for communicating with the field connection controllers, which are RTUs and PLCs, and includes the HMI software running on operator workstations. In smaller SCADA systems, the supervisory computer may be composed of a single PC, in which case the HMI is a part of this computer. In larger SCADA systems, the master station may include several HMIs hosted on client computers, multiple servers for data acquisition, distributed software applications, and disaster recovery sites. To increase the integrity of the system the multiple servers will often be configured in a dual-redundant or hot-standby formation providing continuous control and monitoring in the event of a server malfunction or breakdown.

Remote Terminal Units

Remote terminal units, also known as (RTUs), connect to sensors and actuators in the process, and are networked to the supervisory computer system. RTUs are "intelligent I/O" and often have embedded control capabilities such as ladder logic in order to accomplish boolean logic operations.

Programmable Logic Controllers

Also known as PLCs, these are connected to sensors and actuators in the process, and are networked to the supervisory system in the same way as RTUs. PLCs have more sophisticated embedded control capabilities than RTUs, and are programmed in one or more IEC 61131-3 programming languages. PLCs are often used in place of RTUs as field devices because they are more economical, versatile, flexible and configurable.

Communication Infrastructure

This connects the supervisory computer system to the RTUs and PLCs, and may use industry standard or manufacturer proprietary protocols. Both RTU's and PLC's operate autonomously on the near-real time control of the process, using the last command given from the supervisory system. Failure of the communications network does not necessarily stop the plant process controls, and on resumption of communications, the operator can continue with monitoring and control. Some critical systems will have dual redundant data highways, often cabled via diverse routes.

Human-machine Interface

The human-machine interface (HMI) is the operator window of the supervisory system. It presents plant information to the operating personnel graphically in the form of mimic diagrams, which are a schematic representation of the plant being controlled, and alarm and event logging pages. The HMI is linked to the SCADA supervisory computer to provide live data to drive the mimic diagrams, alarm displays and trending graphs. In many installations the HMI is the graphical user interface for the operator, collects all data from external devices, creates reports, performs alarming, sends notifications, etc.

Mimic diagrams consist of line graphics and schematic symbols to represent process elements, or may consist of digital photographs of the process equipment overlain with animated symbols.

Supervisory operation of the plant is by means of the HMI, with operators issuing commands using mouse pointers, keyboards and touch screens. For example, a symbol of a pump can show the operator that the pump is running, and a flow meter symbol can show how much fluid it is pumping through the pipe. The operator can switch the pump off from the mimic by a mouse click or screen touch. The HMI will show the flow rate of the fluid in the pipe decrease in real time.

The HMI package for a SCADA system typically includes a drawing program that the operators or system maintenance personnel use to change the way these points are represented in the interface. These representations can be as simple as an on-screen traffic light, which represents the state of an actual traffic light in the field, or as complex as a multi-projector display representing the position of all of the elevators in a skyscraper or all of the trains on a railway.

A "historian", is a software service within the HMI which accumulates time-stamped data, events, and alarms in a database which can be queried or used to populate graphic trends in the HMI. The historian is a client that requests data from a data acquisition server.

Alarm Handling

An important part of most SCADA implementations is alarm handling. The system monitors whether certain alarm conditions are satisfied, to determine when an alarm event has occurred. Once an alarm event has been detected, one or more actions are taken (such as the activation of one or more alarm indicators, and perhaps the generation of email or text messages so that management or remote SCADA operators are informed). In many cases, a SCADA operator may have to acknowledge the alarm event; this may deactivate some alarm indicators, whereas other indicators remain active until the alarm conditions are cleared.

Alarm conditions can be explicit—for example, an alarm point is a digital status point that has either the value NORMAL or ALARM that is calculated by a formula based on the values in other analogue and digital points—or implicit: the SCADA system might automatically monitor whether the value in an analogue point lies outside high and low- limit values associated with that point.

Examples of alarm indicators include a siren, a pop-up box on a screen, or a coloured or flashing area on a screen (that might act in a similar way to the "fuel tank empty" light in a car); in each case, the role of the alarm indicator is to draw the operator's attention to the part of the system 'in alarm' so that appropriate action can be taken.

PLC/RTU Programming

"Smart" RTUs, or standard PLCs, are capable of autonomously executing simple logic processes without involving the supervisory computer. They employ standardized control programming languages such as under, IEC 61131-3 (a suite of five programming languages including function block, ladder, structured text, sequence function charts and instruction list), is frequently used to create programs which run on these RTUs and PLCs. Unlike a procedural language like the C or FORTRAN, IEC 61131-3 has minimal training requirements by virtue of resembling historic physical control arrays. This allows SCADA system engineers to perform both the design and implementation of a program to be executed on an RTU or PLC.

A programmable automation controller (PAC) is a compact controller that combines the features and capabilities of a PC-based control system with that of a typical PLC. PACs are deployed in SCADA systems to provide RTU and PLC functions. In many electrical substation SCADA applications, "distributed RTUs" use information processors or station computers to communicate with digital protective relays, PACs, and other devices for I/O, and communicate with the SCADA master in lieu of a traditional RTU.

PLC Commercial Integration

Since about 1998, virtually all major PLC manufacturers have offered integrated HMI/SCADA systems, many of them using open and non-proprietary communications protocols. Numerous specialized third-party HMI/SCADA packages, offering built-in compatibility with most major PLCs, have also entered the market, allowing mechanical engineers, electrical engineers and technicians to configure HMIs themselves, without the need for a custom-made program written by a software programmer. The Remote Terminal Unit (RTU) connects to physical equipment. Typically, an RTU converts the electrical signals from the equipment to digital values. By converting and sending these electrical signals out to equipment the RTU can control equipment.

Communication Infrastructure and Methods

SCADA systems have traditionally used combinations of radio and direct wired connections, although SONET/SDH is also frequently used for large systems such as railways and power stations. The remote management or monitoring function of a SCADA system is often referred to as telemetry. Some users want SCADA data to travel over their pre-established corporate networks or to share the network with other applications. The legacy of the early low-bandwidth protocols remains, though.

SCADA protocols are designed to be very compact. Many are designed to send information only when the master station polls the RTU. Typical legacy SCADA protocols include Modbus RTU, RP-570, Profibus and Conitel. These communication protocols, with the exception of Modbus (Modbus has been made open by Schneider Electric), are all SCADA-vendor specific but are widely adopted and used. Standard protocols are IEC 60870-5-101 or 104, IEC 61850 and DNP3. These communication protocols are standardized and recognized by all major SCADA vendors. Many of these protocols now contain extensions to operate over TCP/IP. Although the use of conventional networking specifications, such as TCP/IP, blurs the line between traditional and industrial networking, they each fulfill fundamentally differing requirements. Network simulation can be used in conjunction with SCADA simulators to perform various 'what-if' analyses.

With increasing security demands (such as North American Electric Reliability Corporation (NERC) and critical infrastructure protection (CIP) in the US), there is increasing use of satellite-based communication. This has the key advantages that the infrastructure can be self-contained (not using circuits from the public telephone system), can have built-in encryption, and can be engineered to the availability and reliability required by the SCADA system operator. Earlier experiences using consumer-grade VSAT were poor. Modern carrier-class systems provide the quality of service required for SCADA.

RTUs and other automatic controller devices were developed before the advent of industry wide standards for interoperability. The result is that developers and their management created a multitude of control protocols. Among the larger vendors, there was also the incentive to create their own protocol to "lock in" their customer base. A list of automation protocols is compiled here.

OLE for process control (OPC) can connect different hardware and software, allowing communication even between devices originally not intended to be part of an industrial network.

Standardisation in the field of mySCADA protocols resulted into the vendor independent protocol called OPC UA (Unified Architecture). OPC UA is starting to be widely adopted among multiple SCADA vendors.

SCADA Architecture Development

SCADA systems have evolved through four generations as follows:

- First generation: "monolithic/Stand Alone": Early SCADA system computing was done by large minicomputers. Common network services did not exist at the time SCADA was developed. Thus SCADA systems were independent systems with no connectivity to other systems. The communication protocols used were strictly proprietary at that time. The first-generation SCADA system redundancy was achieved using a back-up mainframe system connected to all the Remote Terminal Unit sites and was used in the event of failure of the primary mainframe system. Some first generation SCADA systems were developed as "turn key" operations that ran on minicomputers such as the PDP-11 series.

- Second generation: "distributed": SCADA information and command processing was distributed across multiple stations which were connected through a LAN. Information was shared in near real time. Each station was responsible for a particular task, which reduced the cost as compared to First Generation SCADA. The network protocols used were still not standardized. Since these protocols were proprietary, very few people beyond the developers knew enough to determine how secure a SCADA installation was. Security of the SCADA installation was usually overlooked.
- Third generation: "networked": Similar to a distributed architecture, any complex SCADA can be reduced to the simplest components and connected through communication protocols. In the case of a networked design, the system may be spread across more than one LAN network called a process control network (PCN) and separated geographically. Several distributed architecture SCADAs running in parallel, with a single supervisor and historian, could be considered a network architecture. This allows for a more cost-effective solution in very large scale systems.
- Fourth generation: "Web-based": The growth of the internet has led SCADA systems to implement web technologies allowing users to view data, exchange information and control processes from anywhere in the world. The early 2000s saw the proliferation of Web SCADA systems. Web SCADA systems use internet browsers such as Google Chrome and Mozilla Firefox as the graphical user interface (GUI) for the operators HMI. This simplifies the client side installation and enables users to access the system from various platforms with web browsers such as servers, personal computers, laptops, tablets and mobile phones.

Security Issues

SCADA systems that tie together decentralized facilities such as power, oil, gas pipelines, water distribution and wastewater collection systems were designed to be open, robust, and easily operated and repaired, but not necessarily secure. The move from proprietary technologies to more standardized and open solutions together with the increased number of connections between SCADA systems, office networks and the Internet has made them more vulnerable to types of network attacks that are relatively common in computer security. For example, United States Computer Emergency Readiness Team (US-CERT) released a vulnerability advisory warning that unauthenticated users could download sensitive configuration information including password hashes from an Inductive Automation Ignition system utilizing a standard attack type leveraging access to the Tomcat Embedded Web server. Security researcher Jerry Brown submitted a similar advisory regarding a buffer overflow vulnerability in a Wonderware InBatchClient ActiveX control. Both vendors made updates available prior to public vulnerability release. Mitigation recommendations were standard patching practices and requiring VPN access for secure connectivity. Consequently, the security of some SCADA-based systems has come into question as they are seen as potentially vulnerable to cyber attacks.

In particular, security researchers are concerned about:

- The lack of concern about security and authentication in the design, deployment and operation of some existing SCADA networks.
- The belief that SCADA systems have the benefit of security through obscurity through the use of specialized protocols and proprietary interfaces.
- The belief that SCADA networks are secure because they are physically secured.
- The belief that SCADA networks are secure because they are disconnected from the Internet.

SCADA systems are used to control and monitor physical processes, examples of which are transmission of electricity, transportation of gas and oil in pipelines, water distribution, traffic lights, and other systems used as the basis of modern society. The security of these SCADA systems is important because compromise or destruction of these systems would impact multiple areas of society far removed from the original compromise. For example, a blackout caused by a compromised electrical SCADA system would cause financial losses to all the customers that received electricity from that source. How security will affect legacy SCADA and new deployments remains to be seen.

There are many threat vectors to a modern SCADA system. One is the threat of unauthorized access to the control software, whether it is human access or changes induced intentionally or accidentally by virus infections and other software threats residing on the control host machine. Another is the threat of packet access to the network segments hosting SCADA devices. In many cases, the control protocol lacks any form of cryptographic security, allowing an attacker to control a SCADA device by sending commands over a network. In many cases SCADA users have assumed that having a VPN offered sufficient protection, unaware that security can be trivially bypassed with physical access to SCADA-related network jacks and switches. Industrial control vendors suggest approaching SCADA security like Information Security with a defense in depth strategy that leverages common IT practices.

The reliable function of SCADA systems in our modern infrastructure may be crucial to public health and safety. As such, attacks on these systems may directly or indirectly threaten public health and safety. Such an attack has already occurred, carried out on Maroochy Shire Council's sewage control system in Queensland, Australia. Shortly after a contractor installed a SCADA system in January 2000, system components began to function erratically. Pumps did not run when needed and alarms were not reported. More critically, sewage flooded a nearby park and contaminated an open surface-water drainage ditch and flowed 500 meters to a tidal canal. The SCADA system was directing sewage valves to open when the design protocol should have kept them closed. Initially this was believed to be a system bug. Monitoring of the system logs revealed the malfunctions were the result of cyber attacks. Investigators reported 46 separate instances of malicious outside interference before the culprit was identified. The attacks were made by a disgruntled ex-employee of the company that had installed the SCADA system. The ex-employee was hoping to be hired by the utility full-time to maintain the system.

In April 2008, the Commission to Assess the Threat to the United States from Electromagnetic Pulse (EMP) Attack issued a Critical Infrastructures Report which discussed the extreme vulnerability of SCADA systems to an electromagnetic pulse (EMP) event. After testing and analysis, the Commission concluded: "SCADA systems are vulnerable to an EMP event. The large numbers and widespread reliance on such systems by all of the Nation's critical infrastructures represent a systemic threat to their continued operation following an EMP event. Additionally, the necessity to reboot, repair, or replace large numbers of geographically widely dispersed systems will considerably impede the Nation's recovery from such an assault."

Many vendors of SCADA and control products have begun to address the risks posed by unauthorized access by developing lines of specialized industrial firewall and VPN solutions for TCP/IP-based SCADA networks as well as external SCADA monitoring and recording equipment. The International Society of Automation (ISA) started formalizing SCADA security requirements in 2007 with a working group, WG4. WG4 "deals specifically with unique technical requirements, measurements, and other features required to evaluate and assure security resilience and performance of industrial automation and control systems devices".

The increased interest in SCADA vulnerabilities has resulted in vulnerability researchers discovering vulnerabilities in commercial SCADA software and more general offensive SCADA techniques presented to the general security community. In electric and gas utility SCADA systems, the vulnerability of the large installed base of wired and wireless serial communications links is addressed in some cases by applying bump-in-the-wire devices that employ authentication and Advanced Encryption Standard encryption rather than replacing all existing nodes. In June 2010, anti-virus security company VirusBlokAda reported the first detection of malware that attacks SCADA systems (Siemens' WinCC/PCS 7 systems) running on Windows operating systems. The malware is called Stuxnet and uses four zero-day attacks to install a rootkit which in turn logs into the SCADA's database and steals design and control files. The malware is also capable of changing the control system and hiding those changes. The malware was found on 14 systems, the majority of which were located in Iran.

In October 2013 National Geographic released a docudrama titled *American Blackout* which dealt with an imagined large-scale cyber attack on SCADA and the United States' electrical grid.

Programmable Automation Controller

Programmable automation controller (PAC) is a term that is loosely used to describe any type of automation controller that incorporates higher-level instructions. The systems are used in industrial control systems (ICS) for machinery in a wide range of industries, including those involved in critical infrastructure.

A PAC makes it possible to provide more complex instructions to automated equipment, enabling much the same capabilities as PC-based controls in an all-in-one package like a programmable logic controller (PLC).

PLCs were created in the 1960s as an improvement over relay-based systems. Although more advanced than relay, PLCs still functioned by simple ladder logic that resembled the appearance of wiring diagrams of relay systems. In the beginning, PLCs had limited memory, required proprietary terminals and lacked remote I/O (input/output) capabilities. Additional abilities required adding hardware cards. PC-based programming of PLC was introduced in the 1980s and offered greater abilities, more memory and sequential control.

Early PACs came on the scene at the beginning of the 21st century. PACs offered a combination of the abilities and technologies of distributed control systems (DCS) and remote terminal units (RTU) as well as some of the abilities offered by PC control. PACs offered more connectivity options and broader control while maintaining smaller packaging and durability for environmental stresses and shock. With these new improvements, PACs were widely adopted.

Controllers of both types (PAC and PLC) have advanced since their creation. With the increased capabilities of PLC, the differentiating lines between the two have blurred. Higher-end PLCs with increased capabilities are often marketed as PAC.

References

- Discrete-control-on-off, control-systems, automation: automationkingz.com, Retrieved 14 April, 2020
- Richard E. Bellman (December 8, 2015). Adaptive Control Processes: A Guided Tour. Princeton University Press. ISBN 9781400874668
- On-off-control-theory-controller: electrical4u.com, Retrieved 16 March, 2020
- Bose, Bimal K. (June 2009). "The Past, Present, and Future of Power Electronics". IEEE Industrial Electronics Magazine. 3 (2): 9. doi:10.1109/MIE.2009.932709
- Programmable-automation-controller-PAC: whatis.techtarget.com, Retrieved 24 April, 2020
- Bequette, B. Wayne (2003). Process Control: Modeling, Design, and Simulation. Upper Saddle River, New Jersey: Prentice Hall. p. 129. ISBN 978-0-13-353640-9
- What-are-dc-drives-types-of-electrical-dc-drives: electricaltechnology.org, Retrieved 28 June, 2020

4

Automation and Manufacturing

Automated manufacturing systems perform certain operations like processing, inspection, assembly, and material handling. It also includes computer-integrated manufacturing, automatic process control, automated production and assembly line, etc. This chapter has been carefully written to provide an easy understanding of the aspects related to automation and manufacturing.

Automation in Production Systems

Some elements of the firm's production system are likely to be automated, whereas others will be operated manually or clerically. Automation can be defined as a technology concerned with the application of mechanical, electronic, and computer-based systems to operate and control production.

The automated elements of the production system can be separated into two categories: (1) automation of the manufacturing systems in the factory and (2) computerization of the manufacturing support systems. In modern production systems, the two categories overlap to some extent, because the automated manufacturing systems operating on the factory floor are themselves often implemented by computer systems and connected to the computerized manufacturing support systems and management information system operating at the plant and enterprise levels. The term computer-integrated manufacturing is used to indicate this extensive use of computers in production systems.

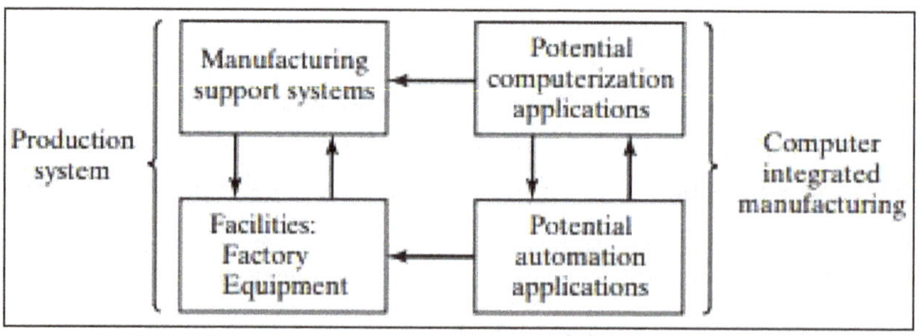

Opportunities of automation and computerization in a production system.

Automated Manufacturing Systems

Automated manufacturing systems operate in the factory on the physical product. They perform operations such as processing, assembly, inspection, or material handling, in some cases accomplishing more than one of these operations in the same system. They are called automated because they perform their operations with a reduced level of human participation compared with the corresponding manual process. In some highly automated systems, there is virtually no human participation. Examples of automated manufacturing systems include:

- Automated machine tools that process parts.
- Transfer lines that perform a series of machining operations.
- Automated assembly systems.
- Manufacturing systems that use industrial robots to perform processing or assembly operations.
- Automatic material handling and storage systems to integrate manufacturing operations.
- Automatic inspection systems for quality control.

Automated manufacturing systems can be classified into three basic types: (1) fixed automation, (2) programmable automation, and (3) flexible automation.

Fixed Automation

Fixed automation is a system in which the sequence of processing (or assembly) operations is fixed by the equipment configuration. Each of the operations in the sequence is usually simple, involving perhaps a plain linear or rotational motion or an uncomplicated combination of the two; for example, the feeding of a rotating spindle. It is the integration and coordination of many such operations into one piece of equipment that makes the system complex. Typical features of fixed automation are:

- High initial investment for custom-engineered equipment.
- High production rates.
- Relatively inflexible in accommodating product variety.

The economic justification for fixed automation is found in products that are produced in very large quantities and at high production rates. The high initial cost of the equipment can be spread over a very large number of units, thus making the unit cost attractive compared with alternative methods of production. Examples of fixed automation include machining transfer lines and automated assembly machines.

Programmable Automation

In programmable automation, the production equipment is designed with the capability to change the sequence of operations to accommodate different product configurations. The operation sequence is controlled by a program, which is a set of instructions coded so that they can be read and interpreted by the system. New programs can be prepared and entered into the equipment to produce new products. Some of the features that characterize programmable automation include:

- High investment in general purpose equipment.
- Lower production rates than fixed automation.
- Flexibility to deal with variations and changes in product configuration.
- Most suitable for batch production.

Programmable automated production systems are used in low- and medium-volume production. The parts or products are typically made in batches. To produce each new batch of a different product, the system must be reprogrammed with the set of machine instructions that correspond to the new product. The physical setup of the machine must also be changed: Tools must be loaded, fixtures must be attached to the machine table, and the required machine settings must be entered. This changeover procedure takes time. Consequently, the typical cycle for a given product includes a period during which the setup and reprogramming takes place, followed by a period in which the batch is produced. Examples of programmable automation include numerically controlled (NC) machine tools, industrial robots, and programmable logic controllers.

Flexible Automation

Flexible automation is an extension of programmable automation. A flexible automated system is capable of producing a variety of parts (or products) with virtually no time lost for changeovers from one part style to the next. There is no lost production time while reprogramming the system and altering the physical setup (tooling, fixtures, machine settings). Consequently, the system can produce various combinations and schedules of parts or products instead of requiring that they be made in batches. What makes flexible automation possible is that the differences between parts processed by the system are not significant. It is a case of soft variety, so that the amount of changeover required between styles is minimal. The features of flexible automation can be summarized as follows:

- High investment for a custom-engineered system.
- Continuous production of variable mixtures of products.
- Medium production rates.
- Flexibility to deal with product design variations.

Examples of flexible automation are the flexible manufacturing systems for performing machining operations that date back to the late 1960s.

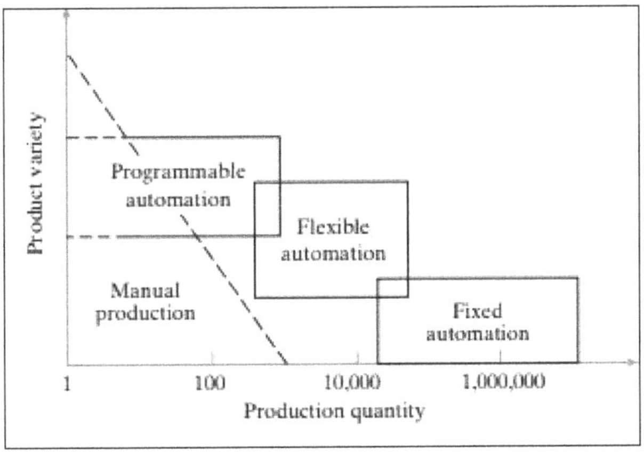

Three types of automation relative to production quantity and product variety.

Computerized Manufacturing Support Systems

Automation of the manufacturing support systems is aimed at reducing the amount of manual and clerical effort in product design, manufacturing planning and control, and the business functions of the firm. Nearly all modern manufacturing support systems are implemented using computer systems. Indeed, computer technology is used to implement automation of the manufacturing systems in the factory as well. The term computer—integrated manufacturing (CIM) denotes the pervasive use of computer systems to design the products, plan the production, control the operations, and perform the various business—related functions needed in a manufacturing firm. True CIM involves integrating all of these functions in one system that operates throughout the enterprise. Other terms are used to identify specific elements of the CIM system. For example, computer-aided design (CAD) denotes the use of computer systems to support the product design function. Computer-aided manufacturing (CAM) denotes the use of computer systems to perform functions related to manufacturing engineering, such as process planning and numerical control part programming. Some computer systems perform both CAD and CAM, and so the term CAD/CAM is used to indicate the integration of the two into one system. Computer—integrated manufacturing includes CAD/CAM, but it also includes the firm's business functions that are related to manufacturing.

Let us attempt to define the relationship between automation and CIM by developing a conceptual model of manufacturing. In a manufacturing firm, the physical production activities that take place in the factory can be distinguished from the information—processing activities, such as product design and production planning, that usually occur in an office environment. The physical activities include all of the processing, assembly, material handling, and inspection operations that are performed on the product in the factory. These operations come in direct contact with the product during manufacture.

Raw materials flow into one end of the factory and finished products flow out the other end. The physical activities take place inside the factory. In our model, the information—processing activities form a ring that surrounds the factory, providing the data and knowledge required to successfully produce the product. These information—processing activities are accomplished to implement the four basic manufacturing support functions identified earlier: (1) business functions, (2) product design, (3) manufacturing planning, and (4) manufacturing control. These four functions form a cycle of events that must accompany the physical production activities but do not directly touch the product.

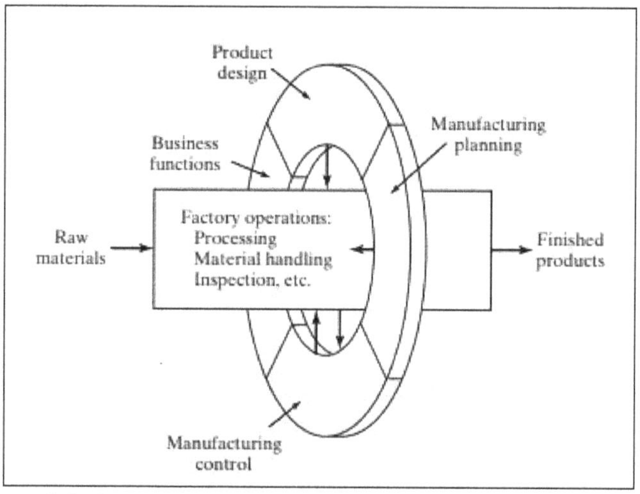

Model of manufacturing showing factory operations and the information—processing activities for manufacturing support.

Reasons for Automating

Companies undertake projects in manufacturing automation and computer-integrated manufacturing for a variety of good reasons. Some of the reasons used to justify automation are the following:

- To increase labor productivity: Automating a manufacturing operation usually increases production rate and labor productivity. This means greater output per hour of labor input.

- To reduce labor cost: Ever-increasing labor cost has been and continues to be the trend in the world's industrialized societies. Consequently, higher investment in automation has become economically justifiable to replace manual operations. Machines are increasingly being substituted for human labor to reduce unit product cost.

- To mitigate the effects of labor shortages: There is a general shortage of labor in many advanced nations, and this has stimulated the development of automated operations as a substitute for labor.

- To reduce or eliminate routine manual and clerical tasks: An argument can be put forth that there is social value in automating operations that are routine, boring, fatiguing, and possibly irksome. Automating such tasks serves a purpose of improving the general level of working conditions.
- To improve worker safety: By automating a given operation and transferring the worker from active participation in the process to a supervisory role, the work is made safer. The safety and physical well-being of the worker has become a national objective with the enactment of the Occupational Safety and Health Act (OSHA) in 1970. This has provided an impetus for automation.
- To improve product quality: Automation not only results in higher production rates than manual operations; it also performs the manufacturing process with greater uniformity and conformity to quality specifications. Reduction of fraction defect rate is one of the chief benefits of automation.
- To reduce manufacturing lead time: Automation helps to reduce the elapsed time between customer order and product delivery, providing a competitive advantage to the manufacturer for future orders. By reducing manufacturing lead time, the manufacturer also reduces work-in-process inventory.
- To accomplish processes that cannot be done manually: Certain operations cannot be accomplished without the aid of a machine. These processes have requirements for precision, miniaturization, or complexity of geometry, that cannot be achieved manually. Examples include certain integrated circuit fabrication operations, rapid prototyping processes based on computer graphics (CAD) models, and the machining of complex, mathematically defined surfaces using computer numerical control. These processes can only be realized by computer controlled systems.
- To avoid the high cost of not automating: There is a significant competitive advantage gained in automating a manufacturing plant. The advantage cannot easily be demonstrated on a company's project authorization form. The benefits of automation often show up in unexpected and intangible ways, such as in improved quality, higher sales, better labor relations, and better company image. Companies that do not automate are likely to find themselves at a competitive disadvantage with their customers, their employees, and the general public.

Automated Production Line

An automated production line is comprised of a series of workstations linked by a transfer system and an electrical control system. Each station performs a specific operation and the product is processed step by step as it moves along the line in a pre-defined production sequence.

A fully automated production line does not need people directly involved in the operation, and all or part of the process of the production is completed by mechanical equipment and automated systems. Therefore, in an automated environment, the tasks of human are more likely to change to system design, adjustment, supervision and monitoring the operation of the system rather than controlling it directly.

There are three types of automation in production: Hard automation (also known as 'fixed automation'), programmable automation and soft automation (also known as 'flexible automation'). The type of automation to utilise is determined by the type of product and volume.

Following the programmed commands, an automated production line is a process that raw materials enter and finished products leave, with little or no human intervention. The fast, stable and accurate production flow contributes to the reduction of production times and the cost of the manufactured products. The use of automated production lines significantly reduces production costs and labour costs, and minimizes human errors, ensuring output consistency and quality. The use of automated production lines frees up people from the drudgery of repetitive tasks – replace human labours in tasks done in dangerous environments, and perform tasks that are beyond human capabilities of size, weight, speed and endurance.

Carefully planned investment in automation can also make good financial sense. A two or three year payback on reduced labour costs is the equivalent of an excellent 30%-50% return on investment.

Automated Assembly Line

Assembly involves the joining together of two or more separate parts to form new entity which may be assembly or subassembly.

Automated assembly refers to the use of mechanized and automated devices to perform the various functions in an assembly line or cell. Automated assembly system performs a sequence of automated operations to combine multiple components in to a single entity which can be a final product or sub assembly.

Automated assembly technology should be considered when the following condition exists:

- High product demand.
- Stable product design.
- The assembly consists of no more than a limited number of components.
- The product is designed for automated assembly.

Automated assembly system involves less investment compared to transfer lines because,

- Work part produced are smaller in size compared to transfer lines.
- Assembly operations do not have the large mechanical forces and power requirement.
- Size is very less compared to transfer lines.

Designs for Automated Assembly

Recommendations and principles that can be applied in product design to facilitate automated assembly.

- Reduce the amount of assembly required: This principle can be realized during design by combining functions within the same part that were previously accomplished by separate components in the product. The use of plastic molded parts to substitute for sheet metal parts is an example of this principle. A more complex geometry molded into a plastic part might replace several metal parts. Although the plastic part may seem to be more costly, the savings-in assembly time probably justify the substitution in many cases.
- Use modular design: In automated assembly, increasing the number of separate assembly steps that are done by a single automated system will result in an increase in the downtime of the system. To reduce this effect, Riley suggests that the design of the product be modular, with perhaps each module requiring a maximum of 12 or 13 parts to be assembled on a single assembly system. Also, the subassembly should be designed around a base part to which other components are added.
- Reduce the number of fasteners required: Instead of using separate screws and nuts, and similar fasteners, design the fastening mechanism into the component design using snap fits and similar features. Also, design the product modules so that several components are fastened simultaneously rather than each component fastened separately.
- Reduce the need for multiple components to lie handled at once: The preferred practice in automated assembly machine design is to separate the operations at different stations rather than to handle and fasten multiple components simultaneously at the same workstation. (It should be noted that robotics technology is causing a rethinking of this practice since robots can be programmed to perform more complex assembly tasks than a single station in a mechanized assembly system.
- Limit the required directions of access: This principle simply means that the number of directions in which new components are added to the existing

subassembly should be minimized. If all of the components can be added vertically from above, this is the ideal situation. Obviously, the design of the subassembly module determines this.

- Require high quality in components: High performance of the automated assembly system requires consistently good quality of the components that are added at each workstation. Poor-quality components cause jams in the feeding and assembly mechanisms which cause downtime in the automated system.
- Implement hopperability: This is a term that is used to identify the ease with which a given component can be fed and oriented reliably for delivery from the parts hopper to the assembly workhead.

Computer Numerical Control

A CNC machine that operates on wood.

Numerical control is the automated control of machining tools (drills, boring tools, lathes) and 3D printers by means of a computer. A CNC machine processes a piece of material (metal, plastic, wood, ceramic, or composite) to meet specifications by following a coded programmed instruction and without a manual operator.

A CNC machine is a motorized maneuverable tool and often a motorized maneuverable platform, which are both controlled by a computer, according to specific input instructions. Instructions are delivered to a CNC machine in the form of a sequential program of machine control instructions such as G-code and then executed. The program can be written by a person or, far more often, generated by graphical computer-aided design (CAD) software. In the case of 3D printers, the part to be printed is "sliced", before the instructions (or the program) is generated. 3D printers also use G-Code.

CNC is a vast improvement over non-computerized machining that must be manually controlled (e.g., using devices such as hand wheels or levers) or mechanically controlled by pre-fabricated pattern guides (cams). In modern CNC systems, the design of a mechanical part and its manufacturing program is highly automated. The part's mechanical dimensions are defined using CAD software and then translated into manufacturing directives by computer-aided manufacturing (CAM) software. The resulting directives are transformed (by "post processor" software) into the specific commands necessary for a particular machine to produce the component, and then are loaded into the CNC machine.

Since any particular component might require the use of a number of different tools – drills, saws, etc. – modern machines often combine multiple tools into a single "cell". In other installations, a number of different machines are used with an external controller and human or robotic operators that move the component from machine to machine. In either case, the series of steps needed to produce any part is highly automated and produces a part that closely matches the original CAD.

Motion is controlling multiple axes, normally at least two (X and Y), and a tool spindle that moves in the Z (depth). The position of the tool is driven by direct-drive stepper motors or servo motors in order to provide highly accurate movements, or in older designs, motors through a series of step-down gears. Open-loop control works as long as the forces are kept small enough and speeds are not too great. On commercial metalworking machines, closed loop controls are standard and required in order to provide the accuracy, speed, and repeatability demanded.

As the controller hardware evolved, the mills themselves also evolved. One change has been to enclose the entire mechanism in a large box as a safety measure, often with additional safety interlocks to ensure the operator is far enough from the working piece for safe operation. Most new CNC systems built today are 100% electronically controlled.

CNC-like systems are used for any process that can be described as movements and operations. These include laser cutting, welding, friction stir welding, ultrasonic welding, flame and plasma cutting, bending, spinning, hole-punching, pinning, gluing, fabric cutting, sewing, tape and fiber placement, routing, picking and placing, and sawing.

Examples of CNC Machines

- Mill: Translates programs consisting of specific numbers and letters to move the spindle (or workpiece) to various locations and depths. Many use G-code. Functions include: face milling, shoulder milling, tapping, drilling and some even offer turning. Today, CNC mills can have 3 to 6 axes. Most CNC mills require placing the workpiece on or in them and must be at least as big as the workpiece, but new 3-axis machines are being produced that are much smaller.
- Lathe: Cuts workpieces while they are rotated. Makes fast, precision cuts, generally using indexable tools and drills. Effective for complicated programs designed to make parts that would be infeasible to make on manual lathes. Similar

control specifications to CNC mills and can often read G-code. Generally have two axes (X and Z), but newer models have more axes, allowing for more advanced jobs to be machined.

- Plasma cutter: Involves cutting a material using a plasma torch. Commonly used to cut steel and other metals, but can be used on a variety of materials. In this process, gas (such as compressed air) is blown at high speed out of a nozzle; at the same time, an electrical arc is formed through that gas from the nozzle to the surface being cut, turning some of that gas to plasma. The plasma is sufficiently hot to melt the material being cut and moves sufficiently fast to blow molten metal away from the cut.
- Electric discharge machining: (EDM), also known as spark machining, spark eroding, burning, die sinking, or wire erosion, is a manufacturing process in which a desired shape is obtained using electrical discharges (sparks). Material is removed from the workpiece by a series of rapidly recurring current discharges between two electrodes, separated by a dielectric fluid and subject to an electric voltage. One of the electrodes is called the tool electrode, or simply the "tool" or "electrode," while the other is called the workpiece electrode, or "workpiece".
- Multi-spindle machine: Type of screw machine used in mass production. Considered to be highly efficient by increasing productivity through automation. Can efficiently cut materials into small pieces while simultaneously utilizing a diversified set of tooling. Multi-spindle machines have multiple spindles on a drum that rotates on a horizontal or vertical axis. The drum contains a drill head which consists of a number of spindles that are mounted on ball bearings and driven by gears. There are two types of attachments for these drill heads, fixed or adjustable, depending on whether the centre distance of the drilling spindle needs to be varied.
- Wire EDM: Also known as wire cutting EDM, wire burning EDM, or traveling wire EDM, this process uses spark erosion to machine or remove material from any electrically conductive material, using a traveling wire electrode. The wire electrode usually consists of brass- or zinc-coated brass material. Wire EDM allows for near 90-degree corners and applies very little pressure on the material. Since the wire is eroded in this process, a wire EDM machine feeds fresh wire from a spool while chopping up the used wire and leaving it in a bin for recycling.
- Sinker EDM: Also called cavity type EDM or volume EDM, a sinker EDM consists of an electrode and workpiece submerged in oil or another dielectric fluid. The electrode and workpiece are connected to a suitable power supply, which generates an electrical potential between the two parts. As the electrode approaches the workpiece, dielectric breakdown occurs in the fluid forming a plasma channel and small spark jumps. Production dies and moulds are often

made with sinker EDM. Some materials, such as soft ferrite materials and epoxy-rich bonded magnetic materials are not compatible with sinker EDM as they are not electrically conductive.

- Water jet cutter: Also known as a "waterjet", is a tool capable of slicing into metal or other materials (such as granite) by using a jet of water at high velocity and pressure, or a mixture of water and an abrasive substance, such as sand. It is often used during fabrication or manufacture of parts for machinery and other devices. Waterjet is the preferred method when the materials being cut are sensitive to the high temperatures generated by other methods. It has found applications in a diverse number of industries from mining to aerospace where it is used for operations such as cutting, shaping, carving, and reaming.

Other CNC Tools

Many other tools have CNC variants, including:

- Drills,
- EDMs,
- Embroidery machines,
- Lathes,
- Milling machine,
- Canned cycle,
- Wood routers,
- Sheet metal works (Turret punch),
- Tube, pipe and wire bending machines,
- Hot-wire foam cutters,
- Plasma cutters,
- Water jet cutters,
- Laser cutting,
- Oxy-fuel,
- Surface grinder,
- Cylindrical grinders,
- 3D printing,
- Induction hardening machines,

- Submerged arc welding,
- Glass cutting,
- CNC router.

Tool/Machine Crashing

In CNC, a "crash" occurs when the machine moves in such a way that is harmful to the machine, tools, or parts being machined, sometimes resulting in bending or breakage of cutting tools, accessory clamps, vises, and fixtures, or causing damage to the machine itself by bending guide rails, breaking drive screws, or causing structural components to crack or deform under strain. A mild crash may not damage the machine or tools, but may damage the part being machined so that it must be scrapped.

Many CNC tools have no inherent sense of the absolute position of the table or tools when turned on. They must be manually "homed" or "zeroed" to have any reference to work from, and these limits are just for figuring out the location of the part to work with it, and are not really any sort of hard motion limit on the mechanism. It is often possible to drive the machine outside the physical bounds of its drive mechanism, resulting in a collision with itself or damage to the drive mechanism. Many machines implement control parameters limiting axis motion past a certain limit in addition to physical limit switches. However, these parameters can often be changed by the operator.

Many CNC tools also do not know anything about their working environment. Machines may have load sensing systems on spindle and axis drives, but some do not. They blindly follow the machining code provided and it is up to an operator to detect if a crash is either occurring or about to occur, and for the operator to manually abort the active process. Machines equipped with load sensors can stop axis or spindle movement in response to an overload condition, but this does not prevent a crash from occurring. It may only limit the damage resulting from the crash. Some crashes may not ever overload any axis or spindle drives.

If the drive system is weaker than the machine structural integrity, then the drive system simply pushes against the obstruction and the drive motors "slip in place". The machine tool may not detect the collision or the slipping, so for example the tool should now be at 210mm on the X axis, but is, in fact, at 32mm where it hit the obstruction and kept slipping. All of the next tool motions will be off by −178mm on the X axis, and all future motions are now invalid, which may result in further collisions with clamps, vises, or the machine itself. This is common in open loop stepper systems, but is not possible in closed loop systems unless mechanical slippage between the motor and drive mechanism has occurred. Instead, in a closed loop system, the machine will continue to attempt to move against the load until either the drive motor goes into an overload condition or a servo motor fails to get to the desired position.

Collision detection and avoidance is possible, through the use of absolute position sensors (optical encoder strips or disks) to verify that motion occurred, or torque sensors or power-draw sensors on the drive system to detect abnormal strain when the machine should just be moving and not cutting, but these are not a common component of most hobby CNC tools.

Instead, most hobby CNC tools simply rely on the assumed accuracy of stepper motors that rotate a specific number of degrees in response to magnetic field changes. It is often assumed the stepper is perfectly accurate and never missteps, so tool position monitoring simply involves counting the number of pulses sent to the stepper over time. An alternate means of stepper position monitoring is usually not available, so crash or slip detection is not possible.

Commercial CNC metalworking machines use closed loop feedback controls for axis movement. In a closed loop system, the controller monitors the actual position of each axis with an absolute or incremental encoder. With proper control programming, this will reduce the possibility of a crash, but it is still up to the operator and programmer to ensure that the machine is operated in a safe manner. However, during the 2000s and 2010s, the software for machining simulation has been maturing rapidly, and it is no longer uncommon for the entire machine tool envelope (including all axes, spindles, chucks, turrets, tool holders, tailstocks, fixtures, clamps, and stock) to be modeled accurately with 3D solid models, which allows the simulation software to predict fairly accurately whether a cycle will involve a crash. Although such simulation is not new, its accuracy and market penetration are changing considerably because of computing advancements.

Numerical Precision and Equipment Backlash

Within the numerical systems of CNC programming it is possible for the code generator to assume that the controlled mechanism is always perfectly accurate, or that precision tolerances are identical for all cutting or movement directions. This is not always a true condition of CNC tools. CNC tools with a large amount of mechanical backlash can still be highly precise if the drive or cutting mechanism is only driven so as to apply cutting force from one direction, and all driving systems are pressed tightly together in that one cutting direction. However a CNC device with high backlash and a dull cutting tool can lead to cutter chatter and possible workpiece gouging. Backlash also affects precision of some operations involving axis movement reversals during cutting, such as the milling of a circle, where axis motion is sinusoidal. However, this can be compensated for if the amount of backlash is precisely known by linear encoders or manual measurement.

The high backlash mechanism itself is not necessarily relied on to be repeatedly precise for the cutting process, but some other reference object or precision surface may be used to zero the mechanism, by tightly applying pressure against the reference and setting that as the zero reference for all following CNC-encoded motions. This is similar to the manual machine tool method of clamping a micrometer onto a reference beam and adjusting the Vernier dial to zero using that object as the reference.

Positioning Control System

In numerical control systems, the position of the tool is defined by a set of instructions called the part program.

Positioning control is handled by means of either an open loop or a closed loop system. In an open loop system, communication takes place in one direction only: from the controller to the motor. In a closed loop system, feedback is provided to the controller so that it can correct for errors in position, velocity, and acceleration, which can arise due to variations in load or temperature. Open loop systems are generally cheaper but less accurate. Stepper motors can be used in both types of systems, while servo motors can only be used in closed systems.

Cartesian Coordinates

The G & M code positions are all based on a three dimensional Cartesian coordinate system. This system is a typical plane often seen in mathematics when graphing. This system is required to map out the machine tool paths and any other kind of actions that need to happen in a specific coordinate. Absolute coordinates are what is generally used more commonly for machines and represent the (0,0,0) point on the plane. This point is set on the stock material in order to give a starting point or "home position" before starting the actual machining.

Automatic Process Control

Automatic process control in continuous production processes is a combination of control engineering and chemical engineering disciplines that uses industrial control systems to achieve a production level of consistency, economy and safety which could not be achieved purely by human manual control. It is implemented widely in industries such as oil refining, pulp and paper manufacturing, chemical processing and power generating plants.

There is a wide range of size, type and complexity, but it enables a small number of operators to manage complex processes to a high degree of consistency. The development of large automatic process control systems was instrumental in enabling the design of large high volume and complex processes, which could not be otherwise economically or safely operated.

The applications can range from controlling the temperature and level of a single process vessel, to a complete chemical processing plant with several thousand control loops.

Development of Modern Process Control Operations

A modern control room where plant information and controls are displayed on computer graphics screens. The operators are seated as they can view and control any part of the process from their screens, whilst retaining a plant overview.

Process control of large industrial plants has evolved through many stages. Initially, control would be from panels local to the process plant. However this required a large manpower resource to attend to these dispersed panels, and there was no overall view of the process. The next logical development was the transmission of all plant measurements to a permanently-manned central control room. Effectively this was the centralisation of all the localised panels, with the advantages of lower manning levels and easier overview of the process. Often the controllers were behind the control room panels, and all automatic and manual control outputs were transmitted back to plant. However, whilst providing a central control focus, this arrangement was inflexible as each control loop had its own controller hardware, and continual operator movement within the control room was required to view different parts of the process.

With the coming of electronic processors and graphic displays it became possible to replace these discrete controllers with computer-based algorithms, hosted on a network of input/output racks with their own control processors. These could be distributed around plant, and communicate with the graphic display in the control room or rooms. The distributed control system was born.

The introduction of DCSs allowed easy interconnection and re-configuration of plant controls such as cascaded loops and interlocks, and easy interfacing with other production computer systems. It enabled sophisticated alarm handling, introduced automatic event logging, removed the need for physical records such as chart recorders, allowed the control racks to be networked and thereby located locally to plant to reduce cabling runs, and provided high level overviews of plant status and production levels.

Hierarchy

The accompanying diagram is a general model which shows functional manufacturing levels in a large process using processor and computer-based control.

Referring to the diagram: Level 0 contains the field devices such as flow and temperature sensors (process value readings - PV), and final control elements (FCE), such as control valves; Level 1 contains the industrialised Input/Output (I/O) modules, and their associated distributed electronic processors; Level 2 contains the supervisory computers, which collate information from processor nodes on the system, and provide the operator control screens; Level 3 is the production control level, which does not directly control the process, but is concerned with monitoring production and monitoring targets; Level 4 is the production scheduling level.

Control Model

To determine the fundamental model for any process, the inputs and outputs of the system are defined differently than for other chemical processes. The balance equations are defined by the control inputs and outputs rather than the material inputs. The control model is a set of equations used to predict the behavior of a system and can help determine what the response to change will be. The state variable (x) is a measurable variable that is a good indicator of the state of the system, such as temperature (energy balance), volume (mass balance) or concentration (component balance). Input variable (u) is a specified variable that commonly include flow rates.

It is important to note that the entering and exiting flows are both considered control inputs. The control input can be classified as a manipulated, disturbance, or unmonitored variable. Parameters (p) are usually a physical limitation and something that is fixed for the system, such as the vessel volume or the viscosity of the material. Output (y) is the metric used to determine the behavior of the system. The control output can be classified as measured, unmeasured, or unmonitored.

Types

Processes can be characterized as batch, continuous, or hybrid. Batch applications require that specific quantities of raw materials be combined in specific ways for particular duration to produce an intermediate or end result. One example is the production of adhesives and glues, which normally require the mixing of raw materials in a heated vessel for a period of time to form a quantity of end product. Other important examples are the production of food, beverages and medicine. Batch processes are generally used to produce a relatively low to intermediate quantity of product per year (a few pounds to millions of pounds).

A continuous physical system is represented through variables that are smooth and uninterrupted in time. The control of the water temperature in a heating jacket, for example, is an example of continuous process control. Some important continuous processes are the production of fuels, chemicals and plastics. Continuous processes in manufacturing are used to produce very large quantities of product per year (millions to billions of pounds). Such controls use feedback such as in the PID controller A PID Controller includes proportional, integrating, and derivative controller functions.

Applications having elements of batch and continuous process control are often called hybrid applications.

Control Loops

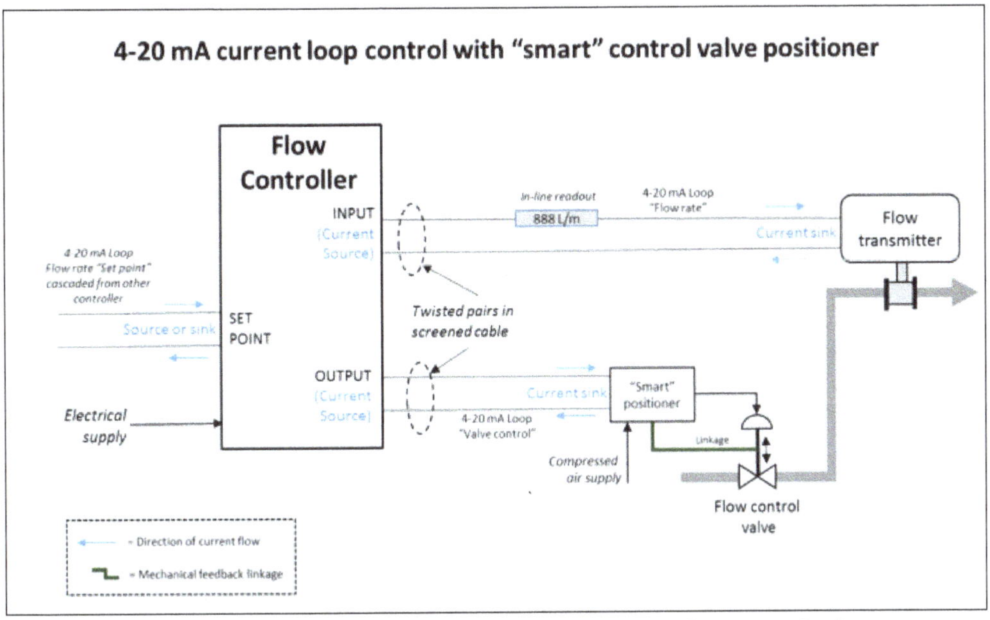

Example of a continuous flow control loop. Signalling is by industry standard 4-20 mA current loops, and a "smart" valve positioner ensures the control valve operates correctly.

The fundamental building block of any industrial control system is the control loop, which controls just one process variable. An example is shown in the accompanying diagram, where the flow rate in a pipe is controlled by a PID controller, assisted by what is effectively a cascaded loop in the form of a valve servo-controller to ensure correct valve positioning.

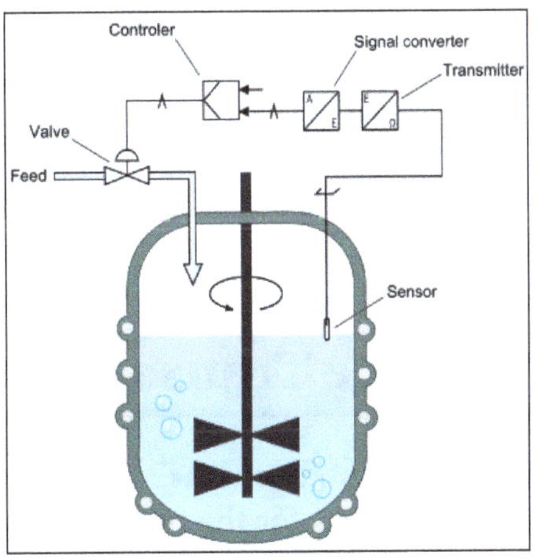

Example of level control system of a continuous stirred-tank reactor. The flow control into the tank would be cascaded off the level control.

Some large systems may have several hundreds or thousands of control loops. In complex processes the loops are interactive, so that the operation of one loop may affect the operation of another. The system diagram for representing control loops is a Piping and instrumentation diagram.

Commonly used controllers are programmable logic controller (PLC), Distributed Control System (DCS) or SCADA.

A further example is shown. If a control valve were used to hold level in a tank, the level controller would compare the equivalent reading of a level sensor to the level setpoint and determine whether more or less valve opening was necessary to keep the level constant. A cascaded flow controller could then calculate the change in the valve position.

Economic Advantages

The economic nature of many products manufactured in batch and continuous processes require highly efficient operation due to thin margins. The competing factor in process control is that products must meet certain specifications in order to be satisfactory. These specifications can come in two forms: a minimum and maximum for a property of the material or product, or a range within which the property must be. All loops are susceptible to disturbances and therefore a buffer must be used on process set points to ensure disturbances do not cause the material or product to go out of specifications. This buffer comes at an economic cost (i.e. additional processing, maintaining elevated or depressed process conditions, etc).

Process efficiency can be enhanced by reducing the margins necessary to ensure product specifications are met. This can be done by improving the control of the process to minimize the effect of disturbances on the process. The efficiency is improved in a two-step method of narrowing the variance and shifting the target. Margins can be narrowed through various process upgrades (i.e. equipment upgrades, enhanced control methods, etc.). Once margins are narrowed, an economic analysis can be done on the process to determine how the set point target is to be shifted. Less conservative process set points lead to increased economic efficiency. Effective process control strategies increase the competitive advantage of manufacturers who employ them.

Enterprise Appliance Transaction Module

An enterprise appliance transaction module (EATM) is a device, typically used in the manufacturing automation marketplace, for the transfer of plant floor equipment and product status to manufacturing execution systems (MES), enterprise resource planning (ERP) systems and the like.

Solutions that deliver manufacturing floor integration have evolved over time. Initially they took the form of custom integrated systems, designed and delivered by System Integrators. These solutions were largely based on separate Commercial off the Shelf (COTS) products that a System Integrator acquired and integrated into the custom solution.

These solutions have evolved into vendor supplied product solutions that no longer require the integration of separate components. EATM products, as they are known today, are available as COTS Appliance products, fully configurable, and not needing any software development or custom integration.

Components Required for EATM Functionality

- Hardware platform: Embedded computer, computer appliance.
- Device communications software: Support for the device protocols from which data will be extracted. Device communications software typically operates through polled or change based protocols that are vendor specific. Data to be extracted is typically organized into related items, and transferred based on a machine status such as Cycle Complete, Job Start, System Downtime Event, Operator Change, etc. Typical protocols; Rockwell Automation CIP, Control-Logix backplane, EtherNet/IP, Siemens Industrial Ethernet, Modbus TCP. There are hundreds of automation device protocols and EATM solutions are typically targeting certain market segments and will be based on automation vendor relationships.
- Enterprise communications software: Software that will enable communications to Enterprise systems. Communications at this level are typically transaction oriented and require data transactions to be sent and acknowledged to ensure the data integrity. Examples include; Relational Database Adapters, Java Message Services (JMS), Oracle Database Interfaces and proprietary interfaces to specific products.
- Transaction application: Software that is configured to watch and collect device variables, formats them into required transactions, and transfer the results securely and reliably to the Enterprise Solutions. The Transaction Application resides between the Device Communications and the Enterprise communications.

Enterprise Solutions

Manufacturing solutions fall into many categories. Overall, the manufacturing environment is portrayed as a three layer Manufacturing Pyramid. At the base, Device Control Systems – Programmable Logic Controller (PLC) and Supervisory Control and Data Acquisition systems (SCADA) perform the process automation functions. A layer above that encompasses Plant Execution Systems that deliver the functions of; Asset

Management, Genealogy, SPC, MES, Order Tracking, Quality Assurance and Scheduling. At the top most level, Enterprise Resource Planning (ERP) systems offer final control over the enterprise and track overall enterprise performance.

It is the job of EATM solutions to act as a bi-directional bridge between field devices and the supervisory control systems. These field devices could be located in a work cell or an assembly or process line. They could be very simple devices, or programmable controllers, machine controls, or PLCs. The upstream business systems could be ANDON and Kanban systems for that line, manufacturing execution systems (MES), and Archival Quality databases.

Lights-out in Manufacturing

Lights-out manufacturing, more often called dark factory is a manufacturing methodology (or philosophy), rather than a specific process.

Factories that run lights out are fully automated and require no human presence on-site. These factories can be considered to run “with the lights off”. Many factories are capable of lights-out production, but very few run exclusively lights-out. Typically, workers are necessary to set up tombstones that hold parts to be manufactured and remove completed parts. As the technology necessary for lights-out production becomes increasingly available, many factories are beginning to use lights-out production between shifts (or as a separate shift) to meet increasing demand or save money. An automatic factory is a place where raw materials enter and finished products leave with little or no human intervention.

One of the earliest descriptions of the automatic factory in fiction was the 1955 short story “Autofac”.

Real-world Examples

- “Lights out” CNC machining: CNC machine tools do not require continuous operator attention, and some models can run unattended. A few machine shops run unattended on nights and weekends.
- Existing “lights-out factories”: FANUC, the Japanese robotics company, has been operating a “lights out” factory for robots since 2001. Robots are building other robots at a rate of about 50 per 24-hour shift and can run unsupervised for as long as 30 days at a time. “Not only is it lights-out,” says Fanuc vice president Gary Zywiol, “we turn off the air conditioning and heat too.” In the Netherlands, Philips uses lights-out manufacturing to produce electric razors, with 128 robots from Adept Technology. The only humans are nine quality assurance workers at the end of the manufacturing process.

In the manufacturing of Integrated circuits using 300mm wafers, the entire manufacturing process is completely automated, with workers only making sure that the process runs without problems and repairing any faulty machinery.

Motivations for Lights-out Factories

Companies that use a largely robotic manufacturing fleet would see more productivity at a lower upkeep cost. Companies incorporating the lights-out factory mindset into floor plans would only need to consider robotic workers. Human labourers would be dispatched to a separate location for tasks such as quality assurance. Using autonomous robotics or optimizing space for a fully industrial force would allow for this increase productivity. James Cook, an applications engineer at Stäubli and colleague of David Arceneaux, the Business Development and Marketing Manager at Stäubli Robotics says robots can help lower building costs by providing for smaller work cells. "Manufacturers can fit a larger number of compact cells in the same space to increase production without adding heating, lighting or cooling to the cost of the building" says Cook.. When considering that floor space is also important in energy considerations, this smaller space will reduce energy consumption by reducing heating costs. Without workers, climate control systems are also unnecessary and smaller layouts will mean less electricity consumption.

Robotics in Manufacturing

Robots in Manufacturing Industries are programmable, mechanical devices used in place of a person to perform dangerous or repetitive tasks with a high degree of accuracy.

Factories started using these machines in the early 1960's to handle some of the more dangerous or mundane tasks that humans didn't want to do. However, they did more than fill unwanted factory jobs; they completed the work with unprecedented speed and precision. Today, robots perform all kinds of tasks and can be classified according to different criteria such as:

- Type of movement,
- Application,
- Architecture,
- Brand,
- Ability to be collaborative.

As labor costs rise and competition for low-wage overseas locations increases, more and more manufacturers are utilizing robot technologies. In fact, 90 percent of all modern robots can be found in factories.

There are six major types of industrial robots used for various tasks:

- Articulated: Articulated robots have rotary joints that allow for a full range of motion. They can perform very precise movements which makes them useful in manufacturing lines where they need to bend in different directions. Multiple arms can be utilized for greater control or to execute multiple tasks at once.

  The main advantage of articulated robots is the flexibility and dexterity. They can move and manipulate a variety of objects while performing small tasks with greater speed and consistency than human workers.

- Cartesian: Also referred to rectilinear or gantry robots, Cartesian robots have three linear joints that move in different axes (X, Y, and Z). The rigid structure of these robots allows for advanced precision and repeatability. They are often used in assembly lines for performing simple movements such as picking up and moving bottles.

  Due to the relatively simple design and mechanic structure, Cartesian robots are fairly inexpensive to make. They can be the cheapest solution for simple pick and place operations or other tasks that do not require extensive movement

- Cylindrical: Just as the name suggests, cylindrical robots have a cylindrical work area. They feature a robotic arm that is connected to a base via single joint, with one more linear joint connecting the arm's links. Basically, these machines feature a single robotic arm that moves up, down, and around a cylindrical pole.

  Cylindrical robots are used for assembly operations, handling, and spot welding. Their function is similar to Cartesian robots, but may be more preferable in some applications due to their ability to move between required points faster.

- Spherical: Spherical robots are similar to, but more complex than, Cartesian or cylindrical robots. They feature a robotic arm connected to a base via a twisting joint, giving the mechanism a spherically shaped work area. This allows them to perform tasks that require movement in a three dimensional space.

  Spherical robots were some of the first industrial robots to be used in manufacturing for construction and other dexterous tasks that require advanced control. Nowadays, though, they are becoming less and less popular as articulated robots are more flexible.

- SCARA: SCARA is an acronym that stands for Selective Compliance Assembly Robot Arm. These robots have arms that behave similarly to a human arm in that the joints allow for both vertical and horizontal movement. However, the "wrist" has limited motion which gives it an advantage for many types of assembly work such as pick and place, kitting, packaging, and other material handling applications.

- Delta: These robots are built from jointed parallelograms connected to a single base, giving them a spider-like appearance. This type of design is optimal for delicate, precise movements that are useful in the food, pharmaceutical, and electronic industries.

Delta robots are used extensively for assembly and other applications that require high-speed repetition. They are able to complete highly repetitive tasks, such as small part assembly, quickly and perfectly each time. This is beneficial not only from an efficiency standpoint, but in terms of health and safety as well. Such tasks have been found to cause musculoskeletal disorders in humans over long periods of time.

Computer Aided Design

Example: 3D CAD model.

Computer-aided design (CAD) is the use of computers (or workstations) to aid in the creation, modification, analysis, or optimization of a design. CAD software is used to increase the productivity of the designer, improve the quality of design, improve communications through documentation, and to create a database for manufacturing. CAD output is often in the form of electronic files for print, machining, or other manufacturing operations. The term CADD (for Computer Aided Design and Drafting) is also used.

Its use in designing electronic systems is known as electronic design automation (EDA). In mechanical design it is known as mechanical design automation (MDA) or computer-aided drafting (CAD), which includes the process of creating a technical drawing with the use of computer software.

CAD software for mechanical design uses either vector-based graphics to depict the objects of traditional drafting, or may also produce raster graphics showing the overall appearance of designed objects. However, it involves more than just shapes. As in the manual drafting of technical and engineering drawings, the output of CAD must convey information, such as materials, processes, dimensions, and tolerances, according to application-specific conventions.

CAD may be used to design curves and figures in two-dimensional (2D) space; or curves, surfaces, and solids in three-dimensional (3D) space.

CAD is an important industrial art extensively used in many applications, including automotive, shipbuilding, and aerospace industries, industrial and architectural design, prosthetics, and many more. CAD is also widely used to produce computer animation for special effects in movies, advertising and technical manuals, often called DCC digital content creation. The modern ubiquity and power of computers means that even perfume bottles and shampoo dispensers are designed using techniques unheard of by engineers of the 1960s. Because of its enormous economic importance, CAD has been a major driving force for research in computational geometry, computer graphics (both hardware and software), and discrete differential geometry.

The design of geometric models for object shapes, in particular, is occasionally called computer-aided geometric design (CAGD).

Starting around the mid 1960s, with the IBM Drafting System, computer-aided design systems began to provide more capability than just an ability to reproduce manual drafting with electronic drafting, the cost-benefit for companies to switch to CAD became apparent. The benefits of CAD systems over manual drafting are the capabilities one often takes for granted from computer systems today; automated generation of bills of materials, auto layout in integrated circuits, interference checking, and many others. Eventually, CAD provided the designer with the ability to perform engineering calculations. During this transition, calculations were still performed either by hand or by those individuals who could run computer programs. CAD was a revolutionary change in the engineering industry, where draftsmen, designers and engineering roles begin to merge. It did not eliminate departments as much as it merged departments and empowered draftsman, designers, and engineers. CAD is an example of the pervasive effect computers were beginning to have on industry. Current computer-aided design software packages range from 2D vector-based drafting systems to 3D solid and surface modelers. Modern CAD packages can also frequently allow rotations in three dimensions, allowing viewing of a designed object from any desired angle, even from the inside looking out. Some CAD software is capable of dynamic mathematical modeling.

CAD technology is used in the design of tools and machinery and in the drafting and design of all types of buildings, from small residential types (houses) to the largest commercial and industrial structures (hospitals and factories).

CAD is mainly used for detailed engineering of 3D models or 2D drawings of physical components, but it is also used throughout the engineering process from conceptual design and layout of products, through strength and dynamic analysis of assemblies to definition of manufacturing methods of components. It can also be used to design objects such as jewelry, furniture, appliances, etc. Furthermore, many CAD applications now offer advanced rendering and animation capabilities so engineers can better visualize their product designs. 4D BIM is a type of virtual construction engineering simulation incorporating time or schedule related information for project management.

CAD has become an especially important technology within the scope of computer-aided technologies, with benefits such as lower product development costs and a greatly shortened design cycle. CAD enables designers to layout and develop work on screen, print it out and save it for future editing, saving time on their drawings.

Uses

Computer-aided design is one of the many tools used by engineers and designers and is used in many ways depending on the profession of the user and the type of software in question.

CAD is one part of the whole digital product development (DPD) activity within the product lifecycle management (PLM) processes, and as such is used together with other tools, which are either integrated modules or stand-alone products, such as:

- Computer-aided engineering (CAE) and finite element analysis (FEA, FEM).
- Computer-aided manufacturing (CAM) including instructions to computer numerical control (CNC) machines.
- Photorealistic rendering and motion simulation.
- Document management and revision control using product data management (PDM).

CAD is also used for the accurate creation of photo simulations that are often required in the preparation of environmental impact reports, in which computer-aided designs of intended buildings are superimposed into photographs of existing environments to represent what that locale will be like, where the proposed facilities are allowed to be built. Potential blockage of view corridors and shadow studies are also frequently analyzed through the use of CAD.

CAD has been proven to be useful to engineers as well. Using four properties which are history, features, parametrization, and high-level constraints. The construction history can be used to look back into the model's personal features and work on the single area rather than the whole model. Parameters and constraints can be used to determine the size, shape, and other properties of the different modeling elements. The features in the CAD system can be used for the variety of tools for measurement such as tensile strength, yield strength, electrical or electromagnetic properties. Also its stress, strain, timing or how the element gets affected in certain temperatures, etc.

Types

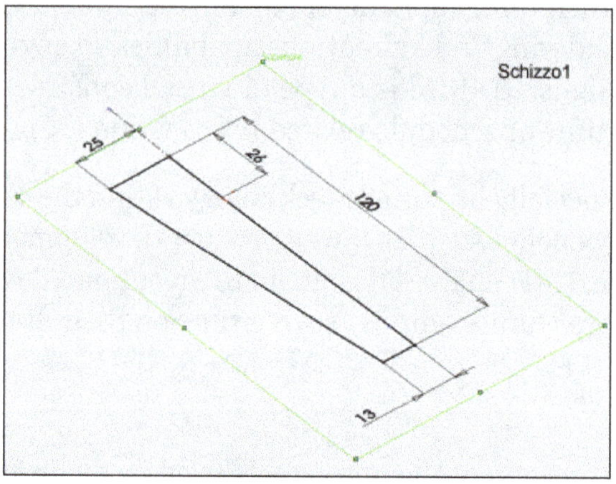

A simple procedure.

There are several different types of CAD, each requiring the operator to think differently about how to use them and design their virtual components in a different manner for each.

There are many producers of the lower-end 2D systems, including a number of free and open-source programs. These provide an approach to the drawing process without all the fuss over scale and placement on the drawing sheet that accompanied hand drafting since these can be adjusted as required during the creation of the final draft.

3D wireframe is basically an extension of 2D drafting (not often used today). Each line has to be manually inserted into the drawing. The final product has no mass properties associated with it and cannot have features directly added to it, such as holes. The operator approaches these in a similar fashion to the 2D systems, although many 3D systems allow using the wireframe model to make the final engineering drawing views.

3D "dumb" solids are created in a way analogous to manipulations of real-world objects (not often used today). Basic three-dimensional geometric forms (prisms, cylinders, spheres, and so on) have solid volumes added or subtracted from them as if assembling or cutting real-world objects. Two-dimensional projected views can easily be generated from the models. Basic 3D solids don't usually include tools to easily allow motion of components, set limits to their motion, or identify interference between components.

There are two types of *3D* solid modeling:

- Parametric modeling allows the operator to use what is referred to as "design intent". The objects and features created are modifiable. Any future modifications can be made by changing how the original part was created. If a feature was intended to be located from the center of the part, the operator should locate it from the center of the model. The feature could be located using any geometric object already available in the part, but this random placement would defeat the design intent. If the operator designs the part as it functions the parametric modeler is able to make changes to the part while maintaining geometric and functional relationships.

- Direct or explicit modeling provide the ability to edit geometry without a history tree. With direct modeling, once a sketch is used to create geometry the sketch is incorporated into the new geometry and the designer just modifies the geometry without needing the original sketch. As with parametric modeling, direct modeling has the ability to include relationships between selected geometry (e.g., tangency, concentricity).

Top end systems offer the capabilities to incorporate more organic, aesthetics and ergonomic features into designs. Freeform surface modeling is often combined with solids to allow the designer to create products that fit the human form and visual requirements as well as they interface with the machine.

Technology

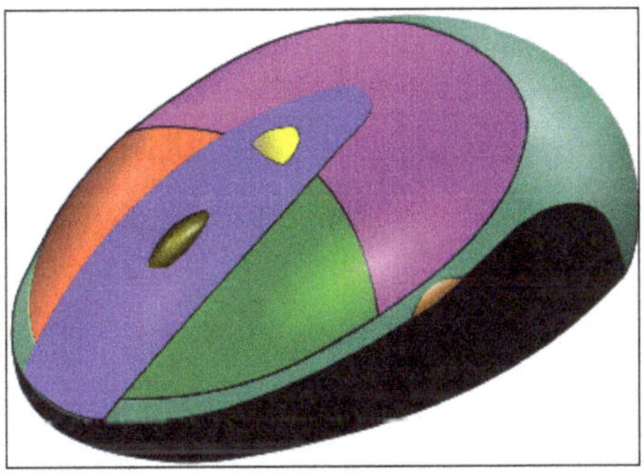

A CAD model of a computer mouse.

Originally software for CAD systems was developed with computer languages such as Fortran, ALGOL but with the advancement of object-oriented programming methods this has radically changed. Typical modern parametric feature-based modeler and freeform surface systems are built around a number of key C modules with their own APIs. A CAD system can be seen as built up from the interaction of a graphical user interface (GUI) with NURBS geometry or boundary representation (B-rep) data via a geometric modeling kernel. A geometry constraint engine may also be employed to manage the associative relationships between geometry, such as wireframe geometry in a sketch or components in an assembly.

Unexpected capabilities of these associative relationships have led to a new form of prototyping called digital prototyping. In contrast to physical prototypes, which entail manufacturing time in the design. That said, CAD models can be generated by a computer after the physical prototype has been scanned using an industrial CT scanning machine. Depending on the nature of the business, digital or physical prototypes can be initially chosen according to specific needs.

Today, CAD systems exist for all the major platforms (Windows, Linux, UNIX and Mac OS X); some packages support multiple platforms.

Currently, no special hardware is required for most CAD software. However, some CAD systems can do graphically and computationally intensive tasks, so a modern graphics card, high speed (and possibly multiple) CPUs and large amounts of RAM may be recommended.

The human-machine interface is generally via a computer mouse but can also be via a pen and digitizing graphics tablet. Manipulation of the view of the model on the screen is also sometimes done with the use of a Spacemouse/SpaceBall. Some systems also support stereoscopic glasses for viewing the 3D model. Technologies which in the past were limited to larger installations or specialist applications have become available to a wide group of users. These include the CAVE or HMDs and interactive devices like motion-sensing technology

References

- Benefits-automated-production-lines: rnaautomation.com, Retrieved 16 June, 2020
- Mahalik, N P (2009). “Processing and packaging automation systems: a review” (PDF). Sens. & Instrumen. Food Qual. 3: 12–25. Retrieved March 23, 2019
- Overview-of-robotics-in-manufacturing: cmtc.com, Retrieved 26 March, 2020
- “Worries about premature industrialisation”. The Economist. Archived from the original on 2017-10-21. Retrieved 2017-10-21
- Computer-Integrated-Manufacturing, Bun-Comp, management: referenceforbusiness.com, Retrieved 15 May, 2020
- “Operational stock of industrial robots at year-end in selected countries” (PDF). Archived from the original (PDF) on 2019-10-11. Retrieved 2019-10-26

5
Industrial Automation Protocols

There are various industrial automation protocols that are used for building automation, power-system automation, vehicular automation, etc. Some of them are DC-BUS, DeviceNet, DNP3, EtherCAT, Modbus, MTConnect, OPC, etc. This chapter discusses the different industrial automation protocols in detail.

AS-Interface

AS-Interface is an intelligent cabling system designed to replace the discrete wires that link a Programmable Controller with sensors and actuators. AS-Interface provides a single cable system that is fast, easy-to-wire and easy-to-operate. Built for the simplest types of devices, AS-Interface is a Master-Slave system where a single AS-Interface Master can exchange I/O data with up to 62 AS-Interface Slave devices. Each Slave can transfer up to four inputs and four outputs on a bus cycle. AS-Interface is a standard hardware and software bus system that brings the reality of truly open systems to the lowest level of the automation business. Your customers are not just asking for connectivity at all levels of the control system, they absolutely expect and require it. Over 11,000,000 AS-Interface devices are now deployed worldwide. If you aren't already in this game, your competitors may already be a step ahead of you.

The first attraction to AS-Interface is its low cost. Cost savings are typically 15 to 40% of other bus systems using traditional cabling methods. Plus the ongoing additional benefits of a bus system ensure enhanced operation and improved maintenance which reduce life cycle costs while improving product quality because of better performance and less downtime.

Second, the sheer blazing speed of AS-Interface attracts a lot of attention. Connected slaves are polled in turn by the AS-Interface Master. A V1 network (31 devices) provides a maximum response time of 5mS per I/O point. In practice, most system have fewer than 31 devices and the cycle times can be significantly shorter. The newer, extended

AS-Interface bus with 62 devices provides a maximum response time of 10msec per I/O point, still much faster than the data acquisition and logic scans of most Programmable Controller systems.

Thirdly, AS-Interface is inherently deterministic and repeatable. Transmissions between the master and slave are predefined and cyclic. The time to set an output is calculable with great accuracy.

And there are other attractions to AS-Interface including a simplicity not found in any other industrial network, device power right on the network cable and no need for Master or Slave Programming. There are no system parameters to set and devices are configured simply by identifying their device ID to the AS-Interface Master. AS-Interface is a very important technology, not only because it addresses the lowest level of a control system architecture, but also because it is backed by Siemens, the worldwide leader in Industrial Control. This gives it an automatic advantage in the marketplace. AS-Interface will be successful for many of the same reasons that systems like DeviceNet have been successful; strong momentum and the backing of a vendor with a huge installed base.

AS-Interface can't be compared to the Sensor bus networks DeviceNet, Profibus, and Interbus-s. All these bus systems are more costly, support more generic communications, provide advanced diagnostic functionality and are highly configurable. These systems support very sophisticated, intelligent devices with tremendous data capabilities. Devices like drive systems, HMIs, barcode readers and the like couldn't find a home on AS-Interfacenterface.

Maximum Number of Nodes	31 in traditional systems, 62 in V2.1.
Number of I/O Points	248 I/O points (4In, 4Out per node) in V1.0, 434 (8In, 3Out per node) in V2.1.
Maximum Cable Length	100m without repeaters, 300m with repeaters.
Power	24VDC
Medium	Unshielded 2 wire cable for both power and data.
Wiring Technology	Insulation displacement for speed and ease of installation.
Topology	Tree Structure
Typical Data Rate	All 31 nodes can be interrogated in 5msec. 10 msec for 62 node systems.
Error Detection	The AS-Interface master automatically detects invalid slave transmissions and schedules repeats.
Configuration	Slaves are automatically configured with a Slave Address by the Master or a Programming Tool. Masters can interrogate the network to identify connected slaves or receive an external list of slave addresses.
Addressing	Each slave gets a specific address set by the AS-Interface Master or Programming Tool.
Master-Slave Operation	A Master polls each slave sequentially and awaits a response.

Hot Swap of Slave Devices	Supported
Multiple Master Operation	Not Supported
Troubleshooting	AS-Interface Masters automatically monitor system voltage levels and slave operation and report this data to the controller.
PLC Programming	No special knowledge of AS-Interface bus or device structure is required to program a PLC to use AS-Interface bus data. AS-Interface data is simply remote IO data to the PLC.
Discrete Device Support	Supported
Analog Device Support	Limited support in V1 and V2.1 using multiple bus cycles (12-bit analog across five bus cycles).

Implementing AS-Interface Quickly and Effectively

Once you've made a decision to add AS-Interface connectivity you have a number of ways to proceed. The choices vary in time-to-market, supportability, resource requirements and price.

- Use an off-the-shelf Serial Gateway: An off-the-shelf serial gateway is the least economical approach with the fastest time-to-market. To implement a serial gateway your product must support a common serial communication protocol such as Modbus, DF/1 or even dumb ASCII. Data from your product will transfer to the gateway at serial baud rates (1200-19200).

  Other than time-to-market (typically a week or two), there are few real advantages to this approach and many negatives. Gateways require an additional footprint, they can be costly, they are generally overkill for a simple device and the gateway vendor appears as the AS-Interface device owner. This option is only attractive if you expect low volume requests for AS-Interface connectivity.

- Use an AS-Interface PCB: If your device uses standard signals, an AS-Interface PCB that accepts your signals and places them on the bus is an alternative solution for quick time-to-market projects. Built to your exact footprint and I/O requirements a custom solution is relatively fast and requires almost none of your resources. Custom solutions offer you the latest in capabilities while giving you absolute control of your development, network presentation and implementation. This approach is great for medium volume applications (i.e. a few hundred units per year).

- Do-It-Yourself: The most costly, lengthy and risky approach is to form an internal effort to build it yourself. While this is admittedly what you'd expect to hear from a company whose business is selling custom networking hardware and software, the facts still speak for themselves. Like most bus implementations there are nuances to the AS-Interface specification that are not readily discernable. For example, the specification provides some very ambiguous language describing how to pick the device ID.

CC-Link Industrial Networks

CC-Link is open industrial network that enables devices from numerous manufacturers to communicate. It is predominantly used in machine, cell or process control applications in manufacturing and production industries, but can also be used in facilities management, process control and building automation.

CC-Link:

- has been approved as IEC 61158 and 61784 Standard for Fieldbus Networks,
- has been approved by ISO (International Organization for Standardization) as 15745-5 International Automation Systems and Integration Standard,
- is a recognized international standard for sensor/actuator networks (SEMI E54.12) by the Semiconductor Equipment Manufacturers Institute,
- is approved as a China National Standard number GB/T 19760-2008 Control and Information Communication Field Network CC-Link Standard and GB/T 20099.4-2006 Chinese Building Automation Standard,
- is an approved Korean National Standard KSBISO15745-5,
- is an approved Taiwan National Standard CNS 15252X6068.

Technical Overview of CLPA Family of Industrial Automation Networks

CC-Link:

- Max.10 Mbit/s network transmission.
- Based on RS485 with networks up to 1.2 km, or extended to 13.2 km with repeaters
- 64 stations per network.
- Refresh time of <3.9mSecs for 65 I/O stations (distance dependent).
- Master/Slave network with Floating Masters and hot swap of stations.

CC-Link Safety:

- Max.10 Mbit/s network transmission.
- Based on CC-Link.
- Can combine Safety devices with standard devices on the same network.
- Compliance with IEC 61508 SIL3 and ISO13849-1 Category 4.

CC-Link LT:

- Max.2.5 Mbit/s network transmission.
- Trunk length up to 500m + 200m drop length.
- 64 stations per network.
- Refresh time of 1.2mSecs for 65 I/O stations (distance dependent).

CC-Link IE Control:

- Dual redundant 1 Gbit/s fibre-optic or Ethernet based network.
- 120 stations per network (Master station: 1, slave station: 119).
- 550 meters between stations.
- Completely deterministic.

CC-Link IE Field:

- 1 Gbit/s Ethernet based network.
- Ethernet physical layer (Category 5e or higher, straight cable with double shielded, STP).
- 121 stations per network (Master station: 1, slave station: 120).
- 100 meters between stations.
- Completely deterministic without using switches.

Data Transmission

Master Frame Transmission Format

Standard Master Frame (Maximum 930 bytes):

F	F	F	A1	A2	871	872	DATA (MAX 918 bytes)	CRC	F	F	F

Master Transmission (Maximum size with Transient Message):

F	F	F	A1	A2	ST1	ST2	RY (MAX 256 bytes)	RWw (MAX 512 bytes)	Message (MAX 150 bytes)	CRC	F	F	F

Master Transmission (Maximum size without Transient Message):

F	F	F	A1	A2	ST1	ST2	RY (MAX 256 bytes)	RWw (MAX 512 bytes)	CRC	F	F	F

Slave Frame Transmission Format

Standard Slave Frame:

F	F	F	A 1	A 2	ST 1	ST2	DATA	CRC	F	F	F

Slave Transmission (Maximum data with Transient Message):

F	F	F	A 1	A 2	ST 1	ST2	RY (MAX 16 bytes)	RWw (MAX 32 bytes)	Message (MAX 34 bytes)	CRC	F	F	F

Slave Transmission (Maximum data without Transient Message):

F	F	F	A 1	A 2	ST 1	ST2	RY (MAX 16 bytes)	RWw (MAX 32 bytes)	CRC	F	F	F

Conformance Testing

All certification testing for CC-Link is carried out by the CLPA. Even though certification is not compulsory, CLPA certification ensures that devices manufactured by suppliers meet the strict technical conformance of CC-Link, including EMI and reaction times. To declare product as CC-Link compatible a vendor needs to send their product to one of the CLPA test lababoratories situated in the USA, China, Korea, Japan and Germany for certification testing.

ControlNet

ControlNet is an open industrial network protocol for industrial automation applications, also known as a fieldbus. ControlNet was earlier supported by ControlNet International, but in 2008 support and management of ControlNet was transferred to ODVA, which now manages all protocols in the Common Industrial Protocol family.

Features which set ControlNet apart from other fieldbuses include the built-in support for fully redundant cables and the fact that communication on ControlNet can be strictly scheduled and highly deterministic. Due to the unique physical layer, common network sniffers such as Wireshark cannot be used to sniff ControlNet packets. Rockwell Automation provides ControlNet Traffic Analyzer software to sniff and analyze ControlNet packets.

Version 1, 1.25 and 1.5

Versions 1 and 1.25 were released in quick succession when ControlNet first launched in 1997. Version 1.5 was released in 1998 and hardware produced for each version variant was typically not compatible. Most installations of ControlNet are version 1.5.

Architecture

Physical Layer

ControlNet cables consist of RG-6 coaxial cable with BNC connectors, though optical fiber is sometimes used for long distances. The network topology is a bus structure with short taps. ControlNet also supports a star topology if used with the appropriate hardware. ControlNet can operate with a single RG-6 coaxial cable bus, or a dual RG-6 coaxial cable bus for cable redundancy. In all cases, the RG-6 should be of quad-shield variety. Maximum cable length without repeaters is 1000m and maximum number of nodes on the bus is 99. However, there is a tradeoff between number of devices on the bus and total cable length. Repeaters can be used to further extend the cable length. The network can support up to 5 repeaters (10 when used for redundant networks). The repeaters do not utilize network node numbers and are available in copper or fiber optic choices.

The physical layer signaling uses Manchester code at 5 Mbit/s.

Link Layer

ControlNet is a scheduled communication network designed for cyclic data exchange. The protocol operates in cycles, known as NUIs, where NUI stands for Network Update Interval. Each NUI has three phases, the first phase is dedicated to scheduled traffic, where all nodes with scheduled data are guaranteed a transmission opportunity. The second phase is dedicated to unscheduled traffic. There is no guarantee that every node will get an opportunity to transmit in every unscheduled phase. The third phase is network maintenance or "guardband". It includes synchronization and a means of determining starting node on the next unscheduled data transfer. Both the scheduled and unscheduled phase use an implicit token ring media access method. The amount of time each NUI consists of is known as the NUT, where NUT stands for Network Update Time. It is configurable from 2 to 100 ms. The default NUT on an unscheduled network is 5 ms.

The maximum size of a scheduled or unscheduled ControlNet data frame is 510 Bytes.

Application Layer

The ControlNet application layer protocol is based on the Common Industrial Protocol (CIP) layer which is also used in DeviceNet and EtherNet/IP.

DC-BUS

DC-BUS is technology for reliable and economical communication over noisy DC or AC power lines. The DC-BUS was originally developed by Yamar Electronics Ltd. together with the DC-BUS Alliance, for low cost sub-networks in vehicles, using the battery lines for in-vehicle data communication. The DC-BUS converts the digital input data into phase modulated signals, protected against errors generated by noise over the power-line. On the receiving side, the received signal is demodulated into the original digital data. Gradually it became a popular means of communication in a plurality of applications within aerospace, automotive, solar energy management and lighting. It is also used as an alternative to RS-232 and RS-485 networks in some cases. The common goal for all these applications is reducing wires, saving space, lowering cost and increasing reliability.

Technology

The DC-BUS is a physical layer that allows transmission of data over the powerline and receiving that signal even if attenuated to the noise level of the powerline. DC-BUS is one of several protocols for power-line communication.

Two technologies implement the DC-BUS: byte oriented and message oriented.

The signaling technology is byte oriented, allowing transfer of a single UART data byte or more over noisy channel (such as the powerline) at bit-rate up to 115.2 kbit/s; each transmitted byte is protected against errors caused by noisy environment. This method operates on a customer selected channel ranging in the HF band. A unique narrow band signaling modulation, based on combination of phase changes is used to transfer each byte. There is no restriction to the number of bytes, therefore, any UART-based standards such as RS-232, RS-485 and LIN-bus can use the DC-BUS as a physical layer. The received data by all the other nodes on that channel has a fixed latency of 3.5 bits. This simplicity of use allows every customer to design his own protocol or use common industry protocols such as LIN. A master slave upper layer protocol is suitable for network operation.

The DCB technology is message oriented, currently operating at speeds up to 1.3Mbit/s. The DCB devices use a selectable narrowband carrier to communicate over the power-line. The DCB gets a message of any length from its host controller and phase modulates it over the power line. The DCB devices have built-in CSMA/CR mechanism for collision detection and arbitration for network operation. The DCB interfaces with its host controller using standard UART or SPI protocols. The use of low level modulated sine wave instead of digital square wave usually between 0 and 12V generates lower unintentional radiation (EMC). The DC-BUS has a sleep mechanism to wake up the DC-BUS devices during sleep mode (low power consumption).

Applications

Automotive

The average modern car contains 1.5 kilometers of copper wire. Wiring plays a significant role in automotive, a passenger car has between 500 and 1500 wires, and these numbers are growing despite the extensive use of CAN and LIN buses. Powerline is common to all the electric modules in a vehicle. The DC-BUS uses these power lines for communication.

- Truck-trailer communication: As part of the SPARC European project conducted by Daimler, two DC-BUS communication networks were used between the truck and the trailer over the existing DC powerline. One network operated at a carrier frequency of 5 MHz for CAN redundant high speed data. The other network operated at 6.5 MHz for trailer backlights activation and control. It saved the existing lights cable between the truck and the trailer. The figure below describes the two networks saving the need for the expensive cable between the truck and the trailer. The experimental InnoTruck used the DC-BUS for its lights control.
- Doors and mirrors control: The DC-BUS can save all the wiring in doors assemblies and its mirrors by using the powerline to control the window, lock and mirror functions (up, down, left, right, blinker, etc.). Only the powerline connects the door to the vehicle body, saving connector and wires. The use of the powerline allows controlling the entertainment unit from the doors, saving the dashboard space.

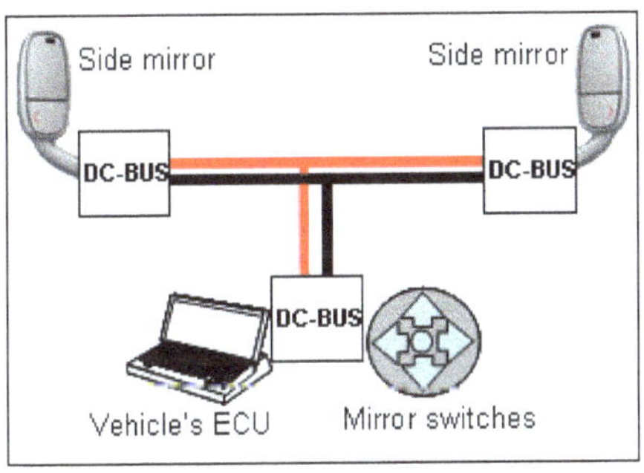

- Steering-wheel control: The figure below presents a steering wheel with a SIG61 device gathering all its switches status and transferring the data over the powerline to the vehicle ECU via the clock-spring saving many wires, reducing the clock spring complexity and cost.

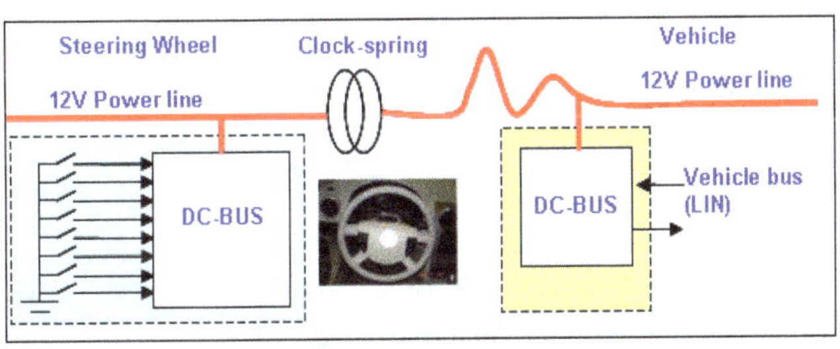

Electrical Vehicle Battery Management

Electrical vehicle (EV) high voltage battery is constructed from serially connected battery power-packs consisting of rechargeable battery cells such as lithiumion. Each cell has to be monitored for proper operation within specified temperature and voltage range. Each power-pack gathers its cell's status and transfers it to the vehicle's battery management system (BMS). Each power-pack is connected at a different voltage reference level. In traditional solution, in order to transfer data, the EV power-packs are connected to the BMS via high voltage data lines, connectors and voltage level shifters. The use of DC-BUS eliminates the need for extra wires, using the powerline as a micro-grid between the power-packs and the BMS. The powerline communication is coupled to the BMS via capacitors that provide galvanic separation between the high voltage powerline and the vehicle's 12/24 V powerline.

Solar Panels Energy Management

Photovoltaic system is constructed from arrays of solar panels and a central energy controller. Communicating between the panels and the central energy management controller allows optimization of each solar panel operating point to increase the entire system energy efficiency and to monitor its proper operation. The DC-BUS constructs a micro-grid over the DC powerline without a need of extra wires. Strings of solar PV panels form entire data network over the high voltage powerline.

Aerospace

Saving wires in aerospace applications is obvious. The DC-BUS was used in the European SCARLETT project for harness reduction purposes.

Home Appliances

Harness of white goods products can be reduced significantly by use of AC/DC-BUS between its modules. The new architecture use only two 24 V DC wires and two 110/220 V AC wires connecting all modules within the white goods products. Yamar's semiconductors use these lines for data communication over noisy power lines; activating loads and reading sensors without need for extra wires.

The 24 V line carries the power required to operate the internal modules as well as for communication between these modules. The 110/220 V wires operate the heavy loads such as compressors and motors.

Lighting

The new generation of illumination devices (LED, CFL bulbs) contains already electronic circuitry. The DC-BUS powerline communication within these bulbs upgrades its functionality as "smart-bulb". Smart-bulb becomes an intelligent sensing and actuation center in addition to its traditional lighting functionality. Each bulb has its own functionality and address allowing controlling its light intensity, color as well as reading the status of its attached sensors. Examples for such sensors are: motion detector, temperature gage, smoke detector etc. Reading data from one smart-bulb can determine the functionality of other smart-bulbs connected to the same powerline.

Standards

- UART: Any UART byte can be transferred over the DC-BUS.
- LIN-bus: The DC-Bus handles the break sequence prior to a LIN message, allowing direct transfer of LIN messages over powerline.
- RS-232: DC-BUS transfers any UART bytes. The DC-BUS communication is half duplex.
- RS-485: DC-BUS transfers any UART bytes. The DC-BUS communication is half duplex.

Development Tools

DC-line Attenuator

The DC-line attenuator is a useful tool for testing the DC-BUS over DC powerline. The attenuator can attenuate the DC-BUS modulated signal between 0 and 60 dB, keeping the DC level of the powerline unaffected. This tool enables measuring the DC-BUS devices bit error rate (BER) in lab.

Limitations

Attenuation of the DC-BUS signal is the main constraint on cable length. For a carrier signal in the HF band, an inductive load such as a motor causes only minor attenuation, but a filtering capacitor represents an extreme attenuation and thus reduces the usable range of the DC-BUS. A DC-BUS device connected close to the battery will have its range further reduced.

Devices which generate significant HF-band signal, such as alternators or microcontroller-based electronics, may cause interference with a poorly-chosen DC-BUS carrier frequency, if they are poorly filtered. Alternately, their filters may significantly attenuate the DC-BUS carrier. Either option reduces the reliability and range of the DC-BUS.

In general, any noise filtering or EMI protection circuitry may significantly reduce the effective range of the DC-BUS, if it attenuates frequencies near the DC-BUS carrier.

DeviceNet

DeviceNet is a network protocol used in the automation industry to interconnect control devices for data exchange. It utilizes the Common Industrial Protocol over a Controller

Area Network media layer and defines an application layer to cover a range of device profiles. Typical applications include information exchange, safety devices, and large I/O control networks.

Architecture

Technical Overview Define the OSI seven-layer architecture model the physical layer, data link layer and application layer.

- Network in addition to the signal, but also including power, self-powered support network function (generally used in small devices, such as photo detectors, limit switches or proximity switches, etc).
- Allows three bit rates: 125 Kbit /s, 250 Kbit /s and 500 Kbit /s, the main trunk under different bit rates (trunk) is inversely proportional to the length and bitrate.
- Flat cable network can be used.
- Single network can have up to 64 nodes, node address (called MAC ID in DeviceNet) by 0–63. Usually the default address for a new out of the box device is 63.
- Duplicate node address detection function.
- Supports master-slave and end-to-end (peer-to-peer) communication architecture, but most of the equipment is operating in the former network architecture.
- Multiple networks allow a single master function.
- Can be used in high-noise environments.

Physical Layer

Nodes are distributed along a DeviceNet network by the means of a trunkline-dropline topology. This topology allows for ease in wiring and access to the network from multiple taps. In addition, nodes can be easily removed and added to reduce production downtime, increase network flexibility, and decrease troubleshooting time. Since the physical layer is optically isolated from the device, communication power and device power can share the same bus (further reducing the complexity of the network and components within).

DeviceNet supports 125 kbit/s, 250 kbit/s and 500 kbit/s data rates. Depending on the chosen cable type, DeviceNet can support communication up to 500 meters (using round, large diameter cable). Typical round cable supports up to 100 meters, while flat-style cable supports up to 380 meters at 125 kbit/s and 75 meters at 500 kbit/s.

Data Link Layer

DeviceNet uses a differential serial bus (Controller Area Network) as its Data Link Layer. Using CAN as a backbone, DeviceNet requires minimal bandwidth to transmit and package messages. In addition, a smaller processor may be selected in the design of the device thanks to the data frame format and the ease at which the processor can parse the data.

CAN Data Frame Format:

```
1 bit => Start of frame

11 bits => Identifier

1 bit => RTR bit

6 bits => Control field

0-8 bytes => Data field

15 bits => CRC sequence

1 bit => CRC delimiter

1 bit => Acknowledge

1 bit => Ack delimiter

7 bits => End of frame

>2 bits => Interframe space
```

Upon transmitting the first packet of data, the Start of Frame bit is sent to synchronize all receivers on the network. The CAN identifier (denoted from 0–63) and RTR bit combine to set priority, at which the data can be accessed or changed. Lower identifiers have priority over higher identifiers. In addition to transmitting this data to other devices, the device also monitors the data sent. This redundancy validates the data transmitted and eliminates simultaneous transmissions. If a node is transmitting at the same time as another node, the node with the lower 11 bit identifier will continue to transmit while the device with the higher 11 bit identifier will stop.

The next six bits contain information for specifying the Control Field. The initial two bits are fixed, while the last four are used to specify length field of the Data Field. The Data Field contains from zero to eight bytes of usable data.

The following data frame is the CRC field. The frame consists of 15 bits to detect frame errors and maintains numerous format delimiters. Due to ease of implementation and immunity to most noisy networks, CAN provides a high level of error checking and fault confinement.

Network

DeviceNet incorporates a connection-based network. A connection must initially be established by either an UCMM (Unconnected Message Manager) or a Group

2 Unconnected Port. From there, Explicit and Implicit messages can be sent and received. Explicit messages are packets of data that generally require a response from another device. Typical messages are configurations or non-time sensitive data collection. Implicit messages are packets of data that are time-critical and generally communicate real-time data over the network. An Explicit Message Connection has to be used to established first before an Implicit Message Connection is made. Once the connection is made, the CAN identifier routes data to the corresponding node.

DNP3

Distributed Network Protocol 3 (DNP3) is a set of communications protocols used between components in process automation systems. Its main use is in utilities such as electric and water companies. Usage in other industries is not common. It was developed for communications between various types of data acquisition and control equipment. It plays a crucial role in SCADA systems, where it is used by SCADA Master Stations, Remote Terminal Units (RTUs), and Intelligent Electronic Devices (IEDs). It is primarily used for communications between a master station and RTUs or IEDs. ICCP, the Inter-Control Center Communications Protocol (a part of IEC 60870-6), is used for inter-master station communications. Competing standards include the older Modbus protocol and the newer IEC 61850 protocol.

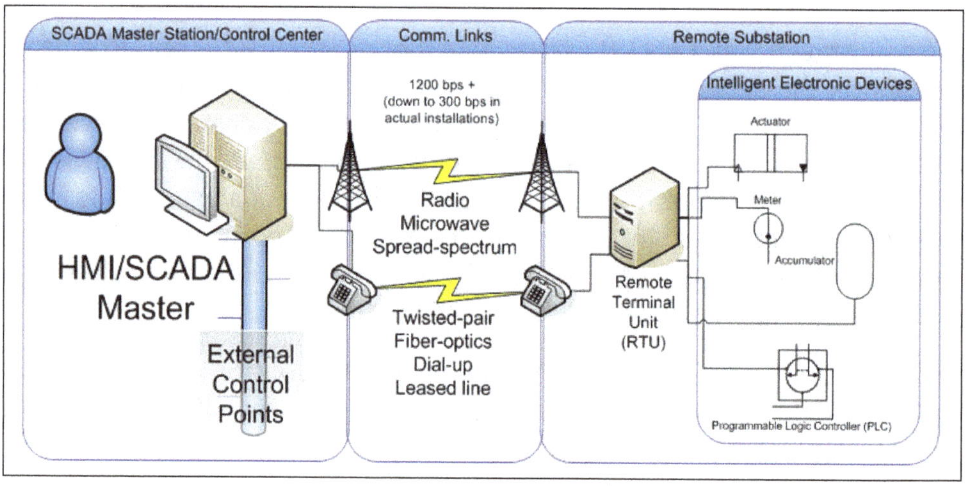

Security

Because smart grid applications generally assume access by third parties to the same physical networks and underlying IP infrastructure of the grid, much work has been done to add Secure Authentication features to the DNP3 protocol. The DNP3 protocol is compliant with IEC 62351-5. Some vendors support encryption via bump-in-the-wire for serial communications or virtual private networks for Internet Protocol-based communications. One of the most popular bump-in-the-wire methods began originally as AGA-12 (American Gas Association) in 2003, later becoming IEEE Std. 1711-2010. This standard was subsequently withdrawn March 27, 2014.

The DNP3 protocol is also referenced in IEEE Std. IEEE 1379-2000, which recommends a set of best practices for implementing modern SCADA Master-RTU/IED communication links. These include not just encryption but other practices that enhance security against well known intrusion methods.

Technical Details

The DNP3 protocol has significant features that make it more robust, efficient, and interoperable than older protocols such as Modbus, at the cost of higher complexity.

In terms of the OSI model for networks, DNP3 specifies a layer 2 protocol. It provides multiplexing, data fragmentation, error checking, link control, prioritization, and layer 2 addressing services for user data. It also defines a Transport function (somewhat similar to the function of layer 4) and an Application Layer (layer 7) that defines functions and generic data types suitable for common SCADA applications. The DNP3 frame strongly resembles, but is not identical to the IEC 60870-5 FT3 frame. It makes heavy use of cyclic redundancy check codes to detect errors.

The improved bandwidth efficiency is accomplished through event oriented data reporting. The Remote Terminal Unit monitors data points and generates events when it determines that the data should be reported (for example, when it changes value). These events are each placed in one of three buffers, associated with "Classes" 1, 2 and 3. In addition to these, Class 0 is defined as the "static" or current status of the monitored data.

The Remote Terminal Unit is initially interrogated with what DNP3 terms an "Integrity Poll" (a combined Read of Class 1, 2, 3 and 0 data). This causes the Remote Terminal Unit to send all buffered events and also all static point data to the Master station. Following this, the Master polls for the event data by reading Class 1, Class 2 or Class 3. The reading of the classes can all be performed together or each class can be read at a different rate, providing a mechanism to create different reporting priorities for the different classes. After an Integrity Poll, only significant data changes are sent. This can result in significantly more responsive data retrieval than polling everything, all the time, irrespective of whether it has changed significantly.

The Remote Terminal Unit can also be configured to spontaneously report Class 1, 2, or 3 data, when it becomes available.

The DNP3 protocol supports time synchronization with an RTU. The DNP Protocol has time stamped variants of all point data objects so that even with infrequent RTU polling, it is still possible to receive enough data to reconstruct a sequence of events of what happened in between the polls. The DNP3 protocol has a substantial library of common point-oriented objects. The focus of this extensive library was to eliminate the need for bit-mapping data over other objects, as is often done in many Modbus installations. For example, floating point number variants are available, so there is no need to map the number on to a pair of 16 bit registers. This improves compatibility and eliminates problems such as endianness.

A Remote Terminal Unit for the DNP3 protocol can be a small, simple embedded device, or it can be a large, complex rack filled with equipment. The DNP User Group has established four levels of subsets of the protocol for RTU compliance. The DNP Users Group has published test procedures for Levels 1 and 2, the simplest implementations.

The protocol is robust, efficient, and compatible with a wide range of equipment, but has become more complex and subtle over time. Increasingly demanding industrial applications are part of the challenge. Also, SCADA concepts are technically simple but field applications that integrate several types of equipment can become complex to set up or troubleshoot due to variances in vendor implementations.

IEEE Standardization

The IEEE adopted DNP3 as IEEE Std 1815-2010 on July 23, 2010. IEEE Std 1815 was co-sponsored by the Transmission and Distribution Committee and Substations Committee of the IEEE Power & Energy Society, with additional input from the DNP Users Group.

In April 2012, the IEEE approved Std 1815-2012 for publication. IEEE Std 1815-2010 has been deprecated. The 2012 version of the standard includes features for Secure Authentication Version 5. The previous version of secure authentication in IEEE 1815-2010 used pre-shared keys only. The new version is capable of using Public Key Infrastructure, and it facilitates remote key changes.

References

- As-interface, technologies: rtautomation.com, Retrieved 30 May, 2020
- Janssen, D.; Büttner, H. (2004), "Real-time Ethernet: the EtherCAT solution", Computing & Control Engineering Journal, 15: 16–21, doi:10.1049/cce:20040104
- Palmer; Shenoi, Sujeet, eds. (23–25 March 2009). Critical Infrastructure Protection III. Third IFIP WG 11. 10 International Conference. Hanover, New Hampshire: Springer. p. 87. ISBN 3-642-04797-1
- Opc-technologies, opc-ua: opcfoundation.org, Retrieved 19 March, 2020
- Lee, Jay; Bagheri, Behrad; Kao, Hung-An (January 2015). "A Cyber-Physical Systems architecture for Industry 4.0-based manufacturing systems". Manufacturing Letters. 3: 18–23. doi:10.1016/j.mfglet.2014.12.001

Permissions

All chapters in this book are published with permission under the Creative Commons Attribution Share Alike License or equivalent. Every chapter published in this book has been scrutinized by our experts. Their significance has been extensively debated. The topics covered herein carry significant information for a comprehensive understanding. They may even be implemented as practical applications or may be referred to as a beginning point for further studies.

We would like to thank the editorial team for lending their expertise to make the book truly unique. They have played a crucial role in the development of this book. Without their invaluable contributions this book wouldn't have been possible. They have made vital efforts to compile up to date information on the varied aspects of this subject to make this book a valuable addition to the collection of many professionals and students.

This book was conceptualized with the vision of imparting up-to-date and integrated information in this field. To ensure the same, a matchless editorial board was set up. Every individual on the board went through rigorous rounds of assessment to prove their worth. After which they invested a large part of their time researching and compiling the most relevant data for our readers.

The editorial board has been involved in producing this book since its inception. They have spent rigorous hours researching and exploring the diverse topics which have resulted in the successful publishing of this book. They have passed on their knowledge of decades through this book. To expedite this challenging task, the publisher supported the team at every step. A small team of assistant editors was also appointed to further simplify the editing procedure and attain best results for the readers.

Apart from the editorial board, the designing team has also invested a significant amount of their time in understanding the subject and creating the most relevant covers. They scrutinized every image to scout for the most suitable representation of the subject and create an appropriate cover for the book.

The publishing team has been an ardent support to the editorial, designing and production team. Their endless efforts to recruit the best for this project, has resulted in the accomplishment of this book. They are a veteran in the field of academics and their pool of knowledge is as vast as their experience in printing. Their expertise and guidance has proved useful at every step. Their uncompromising quality standards have made this book an exceptional effort. Their encouragement from time to time has been an inspiration for everyone.

The publisher and the editorial board hope that this book will prove to be a valuable piece of knowledge for students, practitioners and scholars across the globe.

Index

www.ingramcontent.com/pod-product-compliance
Lightning Source LLC
LaVergne TN
LVHW080628210926
848485LV00003B/228